Clifford Algebras

Daniel Klawitter

Clifford Algebras

Geometric Modelling
and Chain Geometries
with Application in Kinematics

Foreword by Prof. Dr. Gunter Weiss

Dr. Daniel Klawitter
Dresden, Germany

Dissertation TU Dresden, 2014

ISBN 978-3-658-07617-7 ISBN 978-3-658-07618-4 (eBook)
DOI 10.1007/978-3-658-07618-4

The Deutsche Nationalbibliothek lists this publication in the Deutsche Nationalbibliografie; detailed bibliographic data are available in the Internet at http://dnb.d-nb.de.

Library of Congress Control Number: 2014953940

Springer Spektrum
© Springer Fachmedien Wiesbaden 2015
This work is subject to copyright. All rights are reserved by the Publisher, whether the whole or part of the material is concerned, specifically the rights of translation, reprinting, reuse of illustrations, recitation, broadcasting, reproduction on microfilms or in any other physical way, and transmission or information storage and retrieval, electronic adaptation, computer software, or by similar or dissimilar methodology now known or hereafter developed. Exempted from this legal reservation are brief excerpts in connection with reviews or scholarly analysis or material supplied specifically for the purpose of being entered and executed on a computer system, for exclusive use by the purchaser of the work. Duplication of this publication or parts thereof is permitted only under the provisions of the Copyright Law of the Publisher's location, in its current version, and permission for use must always be obtained from Springer. Permissions for use may be obtained through RightsLink at the Copyright Clearance Center. Violations are liable to prosecution under the respective Copyright Law.
The use of general descriptive names, registered names, trademarks, service marks, etc. in this publication does not imply, even in the absence of a specific statement, that such names are exempt from the relevant protective laws and regulations and therefore free for general use.
While the advice and information in this book are believed to be true and accurate at the date of publication, neither the authors nor the editors nor the publisher can accept any legal responsibility for any errors or omissions that may be made. The publisher makes no warranty, express or implied, with respect to the material contained herein.

Printed on acid-free paper

Springer Spektrum is a brand of Springer DE.
Springer DE is part of Springer Science+Business Media.
www.springer-Spektrum.de

To my parents,
Manuela and Burghard

Foreword

What is the right mathematical model to a real phenomenon of our world? Do there exist criteria whether a model can be called elegant as well as practically efficient? These questions will surely be answered differently by e.g. a pure mathematician on the one hand and an engineer on the other. But both will have to start with abstracting real world phenomena to objects of a more or less platonic world. In this idealized world one constructs the " structured geometric image " of real world processes and objects in consideration. Thus, on the way to constructing explicit mathematical models, geometric representations, and geometric reasoning will always play an essential part and provides insight into perhaps not that obvious interrelations and structures.

The idea for this book took its origins from space kinematics and robotics and the standardized way. Its aim is to be understood and handled practically. There are two essentially different ways to model space kinematics, although both ways are based on an Euclidean motion as the core element of the manifold of motions.

Loosely spoken, a seemingly clear model will interpret such a basic element as a point and the manifold of motions as a point manifold. (It turns out, that, via a clever transfer mapping, this point manifold becomes a hyper-quadric in a seven-dimensional real projective space. This hyper-quadric is referred to as Study's quadric.) The second model interprets a motion as a number rather than a point on a line: In order to describe the position of it on that line, one needs a pair of Hamilton quaternions as coordinates, which in addition fulfill conditions. Both models relate to well understood families of models. The first one can be seen e.g. as a higher dimensional analogue to the geometry of lines and screws. Indeed, screws are closely connected to motions. The second model relates to so-called circle geometries.

Their structures are those of a line with coordinates of a ring. (A standard example of such a circle geometry is the so-called (planar) Möbius geometry; here it turns out that the coordinate ring of the corresponding line is even a field, namely the field of complex numbers.)

Both standard models for space kinematics have their merits and can be considered elegant. The first model aims at a visualization of the structured set of motions, where coordinates become inessential. In spite of the transfer between the original set of motions and Study's quadric, the seven-dimensional real projective space is described by coordinates. Now the model induces objects of further research: For example, planar intersections of Study's quadric lead to interesting subsets of motions. The second model aims at a clearly performable calculus and the analytic treatment of forced motions. Here as well, one feels the need for visualization. Such transfer from calculus to geometry establishes a third model and can be derived as follows: Pairs of quaternions are interpreted as elements of a four-dimensional module over the ring of so-called dual numbers, which again are interpreted as points. Points originating from motions are contained by a hypersphere in this four-dimensional module space and, again, planar intersections of this sphere become a matter of interest. But these planar intersections stem from helical motions in the original world.

In this setting, space kinematics and its models are used to point out an obvious fact: The models and model spaces, which are structured for their own, influence the topics of research of the original real world objects and " re-structure " these objects in a specific manner. Therefore, the answer to the introductory question, which model is best, will be rather a matter of taste and pre-knowledge of related models than being based on objectified facts. But still the origin of different models is the object of consideration! So the question about relations between different models and their structures arises. Is is possible to unify these view points?

This last question is the core topic of this book: The author introduces the seemingly universal tool of Clifford Algebras and their

Foreword

group structures to model not only Euclidean kinematics and line geometry, but also circle geometries and their generalizations, so-called " chain geometries " and, to take it even further, also non-Euclidean kinematics and line geometry. Projective and chain geometric properties alternately relate to algebraic properties and concepts. Again, the concept " cross ratio " becomes such an interface between Geometry and Algebra, being on the one hand visualized as quadruplets of points and on the other being a number resp. a ring element as well as responsible for structuring e.g. specific point sets of a line to chains. The presented method allows to connect disciplines of (Applied) Mathematics, that, up to now, are treated independently. Furthermore, it provides an effective tool for solving practical problems of Mechanical Engineering and Robotics, too.

Is this book a Maths book on theoretical and practical applications of Clifford Algebras? Yes, but in its rich details and examples it aims at visualizing the more or less hidden Geometries of the Clifford Algebra calculus, and therefore, it could rightly be called a Geometry book. By its wide range of examples it is profitable to a broad span of readers, be it mechanical engineers or mathematicians. Like me, they surely consider its publication as an absolute stroke of luck.

Wien   Gunter Weiss

Preface

Usually, dual unit quaternions are used to describe Euclidean displacements in three-dimensional kinematics. Engineers widely use this elegant calculus next to a real projective geometric model called Study's quadric. Moreover, there exists a " sphere-model " in a four-dimensional module space over the ring of dual numbers. However, there are no investigations connecting these three models up to now. After recalling Study's quadric, the " sphere-model ", and other models, we present a survey-like introduction to Clifford algebras and their Spin and Pin groups. The advantage of Clifford algebras is that geometric objects and transformation may be represented as algebra elements. Using the so-called sandwich operator, it is possible to apply transformations that are represented by algebra elements directly to geometric objects, also described by algebra elements. We introduce the homogeneous and conformal Clifford algebra model. Furthermore, we show how to model special Cayley-Klein geometries and their isometry groups in a homogeneous Clifford algebra model. As an example, we focus on the homogeneous Clifford algebra model corresponding to line geometry and derive the correspondence between projective transformations of three-dimensional projective space and the Pin group $\text{Pin}_{(3,3,0)}$. Therefore, we introduce the Clifford algebra $\mathcal{Cl}_{(3,3,0)}$, constructed over the quadratic space $\mathbb{R}^{(3,3)}$, and describe how points on Klein's quadric are embedded as null vectors. We discuss how geometric entities that are known from Klein's model, *i.e.*, linear line manifolds can be transferred to this homogeneous Clifford algebra model. All entities known from line geometry occur naturally in this model and can be transformed projectively by the application of the sandwich operator. The action of grade-1 elements corresponds to the action of null polarities on $\mathbb{P}^3(\mathbb{R})$, *i.e.*, correlations that are involutions as the basic elements building up the group of regular projective transformations. It is proven that every regular projective transformation of $\mathbb{P}^3(\mathbb{R})$ can be expressed as the product of six null polarities, *i.e.*, skew-symmetric 4×4 matrices at the most. The results achieved for Klein's quadric may be transferred to any quadric, hence, we present the homogeneous Clifford algebra model corresponding to Lie sphere geometry as an example. Additionally, a new geometric algebra allowing the description of inversions with respect to quadrics in principal position as Pin group is presented. This model serves as a generalization for the conformal geometric algebra and is constructed for dimension two and three in detail. Furthermore, the generalization to arbitrary dimension is shown.

A further focus of the thesis is applying chain geometry to Clifford algebras in order to examine the cross ratio in Clifford algebras. It is well known that the cross ratio of four complex numbers is real if, and only if, they all lie on a Möbius circle, *i.e.*, a circle or a line augmented by a point at infinity. A generalization of the so called Möbius geometry is obtained by using different algebras instead of complex numbers. This leads to a branch of geometry called chain geometry. Chains are subsets of the projective line over a ring which can be parametrised with the cross ratio. Therefore, it is natural to apply this theory to dual quaternions and to examine the kinematic and geometric interpretation. A more general point of view can be achieved by the use of Clifford algebras and Spin groups instead of dual quaternions and dual unit quaternions. After recalling the fundamental chain geometric background, we define the cross ratio for Clifford algebras and their Pin and Spin groups. We present a quadric model corresponding to the dual unit quaternions and homogeneous Clifford algebra models of Klein's, Study's and Lie's quadric where chains that are contained in the grade-1 subspace correspond to conic sections. Moreover, we derive an algebraic biarc construction with the help of contact spaces. Chains of the grade-1 subspace that are in contact at a certain point are parametrized with the cross ratio and correspond to conic sections. Moreover, it has been proven that the connected components of the Pin- and Spin groups define subspaces of chain spaces. Every element contained in a chain defined by three elements of the same connected component of the Pin- or the Spin group is contained in the same connected component of the Pin- or the Spin group. The question for the cross ratio of dual unit quaternions has been answered in detail.

In the last chapter we recall the concept of kinematic mappings. Different kinematic mappings have been found, for example the kinematic mapping of STUDY that maps the group of Euclidean displacements to Study's quadric, or the kinematic mapping of Blaschke and Grünwald. Furthermore, we use the Clifford algebra calculus to unify different kinematic mappings. With this construction it is possible to unify the concept of kinematic mappings for isometry groups of Cayley-Klein geometries and for orthogonal groups $SO(p,q)$. Collineations in any kinematic image and the corresponding Cayley-Klein space can be derived from the homogeneous Clifford algebra model. The kinematic images of Pin- and Spin groups are projective varieties. Due to the fact that projective varieties correspond to ideals, methods of Gröbner basis calculus can be applied to kinematic image spaces. We present kinematic mappings for Euclidean spaces of dimensions two, three and four in detail.

Dresden   Daniel Klawitter

Contents

Foreword   VII

Preface   XI

Introduction   1

1 Models and Representations   5
- 1.1 Description of Displacements 5
  - 1.1.1 Homogeneous Matrices 5
  - 1.1.2 Dual Quaternions 6
  - 1.1.3 Dual Orthogonal Matrices 13
- 1.2 Point Models for Lines and Displacements 15
  - 1.2.1 Klein's Quadric 15
  - 1.2.2 Study's Quadric 21
  - 1.2.3 Study's Sphere 25
- 1.3 Geometric Algebras, Clifford Algebras 26
  - 1.3.1 Definition of a Geometric Algebra 26
  - 1.3.2 Properties of Clifford Algebras 28
  - 1.3.3 Pin and Spin Groups 33
  - 1.3.4 Matrix Representation of Clifford Algebras . . 36
  - 1.3.5 Linear Transformation of the Vector Space . . 37
- 1.4 The Homogeneous Model 38
  - 1.4.1 The Construction 38
  - 1.4.2 Exterior Algebra 39
  - 1.4.3 Homogeneous Model via projective Grassmann Algebra 41
- 1.5 The Conformal Model 42
  - 1.5.1 Construction of Conformal Geometric Algebra 43
  - 1.5.2 Blades in CGA 45
  - 1.5.3 Conformal Transformations 48

1.6	A Clifford algebraic Approach to Line Geometry . . .	50
	1.6.1 Collineations and Correlations in the Image Space	51
	1.6.2 Algebra Representation of Linear Line Manifolds	53
	1.6.3 Transformations	55
	1.6.4 Collineations as Spin Group	61
	1.6.5 Correlations as Pin Group	67
	1.6.6 Singular projective Transformations	72
1.7	A Clifford algebraic Approach to Lie Sphere Geometry	72
	1.7.1 Lie's Quadric	73
	1.7.2 The homogeneous Clifford Algebra Model corresponding to Lie Sphere Geometry	74
1.8	A Clifford algebraic Approach to Study's Quadric . . .	76
1.9	A Clifford algebraic Approach to Study's Sphere . . .	77
1.10	Quadric Geometric Algebra	78
	1.10.1 The Embedding	78
	1.10.2 Geometric Entities	80
	1.10.3 Transformations	84
	1.10.4 Effect on Lines and Points	87
	1.10.5 Subgroups	89
	1.10.6 Generalization to higher Dimensions	94

2 Chain Geometry over Clifford Algebras   101

2.1	Chain Geometry	101
	2.1.1 Distance Spaces	101
	2.1.2 The projective Line over an $\mathcal{L}$-algebra	102
	2.1.3 The Projective Linear Group PGL($\mathcal{R}$, 2)	104
	2.1.4 The projective Line over a Subring	105
2.2	Chain Geometry as Incidence Geometry	106
	2.2.1 Definition of a Cross Ratio	109
2.3	Quadric Chain Spaces	114
2.4	Real Benz Planes	116
2.5	Jordan-Systems	118
2.6	Contact Spaces	121
2.7	Chain Geometries over Clifford Algebras	123
	2.7.1 Grade-1 Subspace	124
	2.7.2 Pin and Spin Groups	126
2.8	Quadric Chain Geometry	131
	2.8.1 Klein's Quadric	131

		2.8.2 Lie's Quadric 133

         2.8.2   Lie's Quadric   133
         2.8.3   Study's Quadric   135
         2.8.4   Study's Sphere   136
  2.9   Quadric Chain Spaces for certain Spin Groups   139
         2.9.1   A Quadric Model for dual unit Quaternions . .   139
         2.9.2   Other possible Quadric Models   141
  2.10   Cross Ratio of dual unit Quaternions   145
         2.10.1 Subspaces on Study's Quadric and Sub Chain
                 Geometries   150
         2.10.2 Application to line-symmetric Displacements .   151
         2.10.3 Dual Quaternion Cross Ratio and Conics on
                 Study's Quadric   156
  2.11   Chains of Geometric Entities   161
         2.11.1 Quadric Geometric Algebra   161
  2.12   Biarc Construction   165
         2.12.1 Biarcs as touching Chains   168
         2.12.2 Biarcs on Quadrics in three-dimensional Space   169
         2.12.3 Klein's Quadric 176
         2.12.4 Biarcs on the Dual Sphere   177

3 Kinematic Mappings for Spin Groups   **181**
  3.1   Cayley-Klein Geometries and the homogeneous Model   181
         3.1.1   Cayley-Klein Spaces   182
         3.1.2   A homogeneous Model for Euclidean Geometry   184
  3.2   Kinematic Mappings   185
         3.2.1   Study's kinematic Mapping   185
         3.2.2   A Mapping for planar Displacements   185
  3.3   Kinematic Mappings via Clifford Algebras   187
         3.3.1   Study's Mapping via Clifford Algebra   187
         3.3.2   Blaschke's and Grünwald's Mapping via Clifford Algebra 190
  3.4   Kinematic mappings for other Cayley-Klein Spaces . .   192
         3.4.1   Two-dimensional Cayley-Klein Spaces   192
         3.4.2   Three-dimensional Cayley-Klein Spaces   193
         3.4.3   Higher dimensional kinematic Mappings   195
  3.5   Projective Varieties via kinematic Algebra Elements .   197

Conclusion   **201**

Index	**203**
List of Symbols	**205**
Acknowledgment	**209**
Bibliography	**211**

Introduction

Euclidean displacements in three-dimensional kinematics are usually described by dual unit quaternions. Engineers widely use this elegant calculus aside a real projective geometric model called Study's quadric. Moreover, there exists a " sphere-model " in a four-dimensional module space over the ring of dual numbers. Seemingly there are no investigations to connect these three models up to now and this was the first motivation for the present research, which turned out to lead to a far more general view. One theme of interest was thereby to find interpretations of different projective closures of the two ring based models and the real projective model.

The cross ratio is a fundamental invariant of projective geometry. It is well known that the cross ratio of four complex numbers is real if, and only if, they all lie on a Möbius circle, *i.e.*, a circle or a line augmented by a point at infinity. A generalization of the so-called Möbius geometry is obtained by using different algebras instead of the complex numbers. A cross ratio for dual quaternions was defined in [58] by

$$cr(a,b,c,d) := (a-b) \cdot (b-c)^{-1} \cdot (c-d) \cdot (d-a)^{-1}.$$

With this definition a lot of questions arise. What is the cross ratio of dual unit quaternions and how can it be interpreted geometrically and kinematically? Under which preliminaries is the cross ratio of four dual unit quaternions a real or a dual number? Are there points on a conic section on Study's quadric corresponding to dual unit quaternions with a real or dual cross ratio?

Kinematic and line geometric models that are used today date back to E. STUDY [61] and F. KLEIN [42]. These models find application in modern treatises, see for example [13], [38]. Especially line geometry experienced a renaissance through computational technology, see [56]. Different representations of Euclidean displacements are well-known. Chapter 1 deals with these models. Study's quadric, the " sphere-model " and other models are described. Furthermore, we present a survey-like introduction to Clifford algebras and their Spin and Pin groups. The advantage of Clifford algebras is that geometric objects and transformation may be represented as algebra elements. With the so-called sandwich operator it is possible to apply transformations that are represented by algebra elements directly to geometric object also described by algebra elements. We introduce the homogeneous and conformal Clifford algebra model. Furthermore, we show how to model special Cayley-Klein geometries and their isometry groups in homogeneous Clifford algebra models. Therefore, we focus on the homogeneous Clifford algebra model corresponding to line geometry and derive the relationship between projective transformations of three-dimensional projective space and the Pin group $\text{Pin}_{(3,3,0)}$.

Homogeneous Coordinates for lines of three-dimensional projective space $\mathbb{P}^3(\mathbb{R})$ were introduced by Plücker, see [7]. The Plücker coordinates of all lines form a quadric in five-dimensional projective space, the so-called Klein quadric denoted by $M_2^4 \subset \mathbb{P}^5(\mathbb{R})$, see [66]. The group of regular projective transformations of $\mathbb{P}^3(\mathbb{R})$ is isomorphic to the group of projective automorphisms of Klein's quadric M_2^4, see [56]. Moreover, the group of automorphic collineations of Klein's quadric is the isometry group of the Cayley-Klein space given by $\mathbb{P}^5(\mathbb{R})$ together with absolute figure M_2^4. This isometry group corresponds to the Pin group of a special homogeneous Clifford algebra model, see [28]. Therefore, we introduce the Clifford algebra $\mathcal{Cl}_{(3,3,0)}$ constructed over the quadratic space $\mathbb{R}^{(3,3)}$ and describe how points on Klein's quadric are embedded as null vectors. We discuss how geometric entities that are known from Klein's model can be transferred to this homogeneous Clifford algebra model. The action of grade-1 elements corresponds to the action of null polarities on $\mathbb{P}^3(\mathbb{R})$. Moreover, we prove that every regular projective transformation of $\mathbb{P}^3(\mathbb{R})$ can be expressed as the product of at most six null polarities, i.e., skew-symmetric 4×4 matrices. The results that were achieved for Klein's quadric may be transferred to any quadric, and therefore, we give the homogeneous Clifford algebra model corresponding to Lie sphere geometry as example. Furthermore, a new geometric algebra which allows the description of inversions with respect to quadrics in principal position as Pin group is presented. This new model is constructed for dimension two and three in detail. Furthermore, the generalization to arbitrary dimension is presented.

In chapter 2 we give the answer to the question for a cross ratio of four dual unit quaternions. This leads to a branch of geometry called chain geometry. Chains are subsets of the projective line over a ring that can be parametrised with the cross ratio. The origins of chain geometry can be found in BENZ [5]. BENZ dealt with two-dimensional algebras over the real numbers, *i.e.*, the double numbers or split-complex numbers $\mathbb{A}$, the complex numbers $\mathbb{C}$, and the dual numbers $\mathbb{D}$. He classified the chains in these three algebras. More recent results can be found in [11] or [33]. Chain geometry is an essential tool for this work. The connection between quadric models and chain geometries over kinematic algebras was found by H. HOTJE, see [37]. A. BLUNCK and M. STROPPEL generalized this approach to so-called Klingenberg chain spaces, see [9] and [12]. Therefore, it is natural to apply this theory to dual quaternions and to examine the kinematic and geometric interpretation. The cross ratio of four dual unit quaternions corresponding to line-symmetric motions with respect to a hyperboloid of revolution

$$\frac{x^2}{a^2} + \frac{y^2}{b^2} - \frac{z^2}{c^2} = 1, \text{ with } a = b,$$

is real-valued, see [58]. However, there is no explanation in general. To answer the question for the cross ratio in all generality is a declared aim of this research. A more general point of view can be achieved with the use of Clifford algebras and Spin groups instead of dual quaternions and dual unit quaternions. The study of the cross ratio of four dual unit quaternions can be translated to the study of the cross ratio of four Spin group elements in a special Clifford algebra. Therefore, we study the cross ratio in the more general Clifford algebra setting. Clifford algebras and especially homogeneous and conformal models allow an elegant description of the group action on geometric entities. There is a lot of literature on these topics see [19, 27, 28] or [45] to mention just a few. With the help of this framework more general properties can be derived, too. Within the Clifford algebra context Spin groups corresponding to more general transformation groups can be discussed. The theory of chain geometry can be applied to every Clifford algebra and its Pin or Spin group. In chapter 2 we give the fundamental chain geometric background and define the cross ratio for Clifford algebras and their Pin and Spin groups. Furthermore, we present a quadric model corresponding to the dual unit quaternions and homogeneous Clifford algebra models whose Pin group corresponds to the group of automorphic collineations of Klein's, Study's and Lie's quadric. Moreover, we derive an algebraic biarc construction with the help of contact spaces. Therefore, the parametrisation of chains with the cross ratio is used to model conic sections on arbitrary quadrics. Biarcs on quadrics are well-known, see [56, 64]. By means of contact relations between chains it is possible to transfer this construction to an algebraic setting. The advantage of this method is that the biarc construction is independent of the model.

In the last chapter we recall the concept of kinematic mappings. Different kinematic mappings have been found, for example the kinematic mapping of STUDY [61] that maps the group of Euclidean displacements to Study's quadric, or the kinematic mapping of Blaschke [6] and Grünwald [26]. Furthermore, we use the Clifford algebra framework to unify different kinematic mappings. We derive a method to map every Pin or Spin group element to a point on certain pseudo-algebraic variety in projective space. The dimension of the projective space depends of the dimension of the Clifford algebra. We define Cayley-Klein spaces and show how to present them in a homogeneous Clifford algebra model. With this concept it is possible to unify the concept of a kinematic mapping for the displacement group of Cayley-Klein geometries and for orthogonal groups $SO(p,q)$. We present kinematic mappings for Euclidean spaces of dimensions two, three and four in detail. Projective varieties correspond to ideals, and therefore, the theory of Gröbner basis calculus can be applied to image spaces.

1 Models and Representations

In this chapter we give a survey on methods that are used nowadays to describe spatial displacements and other transformations. Therefore, point models for lines and spatial displacements in three-dimensional Euclidean space are presented. Moreover, different methods to describe displacements are examined. We review old concepts such as dual numbers, quaternions, and dual quaternions. Furthermore, we give a short review about homogeneous coordinates and how to " linearize " displacements. After that we consider Clifford algebras and special Clifford algebra models. These models are a current field of research.

1.1 Description of Displacements

1.1.1 Homogeneous Matrices

To describe Euclidean transformations we use a familiar trick. We consider the three-dimensional Euclidean space as affine part of the projective space $\mathbb{P}(\mathbb{R} \times \mathbb{R}^3)$ see [31]. The first coordinate is defined as homogeneous factor and we denote the projective space by $\mathbb{P}^3(\mathbb{R})$. Points in $\mathbb{P}^3(\mathbb{R})$ are described by *homogeneous coordinates* $(p_0, p_1, p_2, p_3)^\mathrm{T}$. Note that two homogeneous coordinate vectors x, y describe the same point if $x = f \cdot y$, $f \in \mathbb{R}\setminus\{0\}$. The factor f is called the *homogeneous factor*. A point $p \in \mathbb{P}^3(\mathbb{R})$, $p \neq (0,0,0,0)$ is called *proper* if $p_0 \neq 0$ else the point is called *ideal*. Proper points correspond to affine points with inhomogeneous coordinates. Hence, the projective space without the hyperplane given by $x_0 = 0$ can be identified with the affine space. We can step between the homogeneous and the inhomogeneous representation by $(p_0, p_1, p_2, p_3)^\mathrm{T} \mapsto \left(\frac{p_1}{p_0}, \frac{p_2}{p_0}, \frac{p_3}{p_0}\right)^\mathrm{T}$.

This allows us to describe Euclidean displacements with help of linear transformations in $\mathbb{P}(1 \times \mathbb{R}^3)$ as

$$p' = \begin{pmatrix} 1 & 0^T \\ t & A \end{pmatrix} \cdot p, \quad A \in \mathrm{SO}(3),\ t \in \mathbb{R}^3,\ p = (1, x_1, x_2, x_3)^T. \qquad (1.1)$$

The group of Euclidean displacements that can be represented with Eq. (1.1) is in the focus of this work. In the following we review representations of this group.

1.1.2 Dual Quaternions

Dual quaternions are quaternions with dual number entries. Therefore, we introduce quaternions and dual numbers first.

Quaternions Quaternions $\mathbb{H}$, introduced by W.R. HAMILTON in 1843, constitute an elegant tool for the representation of rotations in three- and four-dimensional Euclidean case, see [49]. Furthermore, quaternions form a skew field, *i.e.*, multiplication is not commutative. Unit quaternions are a double cover of the group SO(3). A general quaternion has the form

$$q = a + b\mathbf{i} + c\mathbf{j} + d\mathbf{k}, \quad a, b, c, d \in \mathbb{R}, \qquad (1.2)$$

where $\mathbf{i}, \mathbf{j}$, and $\mathbf{k}$ are the *quaternion units* with

$$\mathbf{i}^2 = \mathbf{j}^2 = \mathbf{k}^2 = \mathbf{ijk} = -1. \qquad (1.3)$$

Definition 1.1. *Let q be a quaternion as in Eq. (1.2). Then, a is called the* scalar part *and $(b, c, d)^T$ is called the* vector part. *A quaternion with vanishing scalar part is called a* vectorial quaternion.

Quaternions form a four-dimensional vector space over the real numbers. Addition is defined component-wise by

$$q_1 + q_2 = (a_1 + a_2) + (b_1 + b_2)\mathbf{i} + (c_1 + c_2)\mathbf{j} + (d_1 + d_2)\mathbf{k}.$$

The multiplication rules for the quaternion units (1.3) are summarized in the following scheme:

$\cdot$	i	j	k
i	-1	k	$-$j
j	$-$k	-1	i
k	j	$-$i	-1

1.1 Description of Displacements

This scheme allows to extend the concept of quaternion multiplication to general quaternions. The real numbers form the *center* of the quaternions, *i.e.*, real numbers commute with all other quaternions. Later, in section 1.3 we study algebras and we see that the quaternions are an example of an $\mathbb{R}$-algebra.

Definition 1.2. *The* anti-involution

$$* : \mathbb{H} \to \mathbb{H}, \quad a + b\mathbf{i} + c\mathbf{j} + d\mathbf{k} = q \mapsto q^* = a - b\mathbf{i} - c\mathbf{j} - d\mathbf{k}$$

is called quaternion conjugation. The norm of a quaternion is defined by

$$\|q\| = \sqrt{qq^*} = \sqrt{a^2 + b^2 + c^2 + d^2}.$$

Every quaternion $q \neq 0$ has an *inverse quaternion* that can be calculated by

$$q^{-1} = \frac{1}{\|q\|^2} q^* = \frac{q^*}{qq^*}.$$

Unit quaternions are quaternions with norm equal to 1. With respect to quaternion multiplication unit quaternions form a group. Every unit quaternion can be represented by

$$q = \cos\varphi + d\sin\varphi \text{ with } d = (d_1\mathbf{i} + d_2\mathbf{j} + d_3\mathbf{k}),$$

where d is an unit vector and $\varphi \in \mathbb{R}$. Moreover, unit quaternions can be used to describe rotations in three-dimensional Euclidean space $\mathbb{E}^3$. This can be realized with the so called *sandwich operator* $x \mapsto qxq^*$ where q is a unit quaternion and the coordinate vector x of a point is considered as a vectorial quaternion. Therefore, we follow [38] and apply the sandwich operator to the standard basis vectors of $\mathbb{R}^3$ written as vectorial quaternions. This means the basis vectors e_1 belonging to the x-component is expressed by the quaternion $x = \mathbf{i}$ and the effect of the sandwich operator results in

$$\begin{aligned}
q\mathbf{i}q^* &= (a + b\mathbf{i} + c\mathbf{j} + d\mathbf{k})\mathbf{j}(a - b\mathbf{i} - c\mathbf{j} - d\mathbf{k}) \\
&= (a\mathbf{i} - b - c\mathbf{k} + d\mathbf{j})\mathbf{j}(a - b\mathbf{i} - c\mathbf{j} - d\mathbf{k}) \\
&= (a^2 + b^2 - c^2 - d^2)\mathbf{i} + 2(bc + ad)\mathbf{j} + 2(bd - ac)\mathbf{k}.
\end{aligned}$$

In the same way we see the action of this operator on the vectorial quaternions **j** and **k**

$$q\mathbf{j}q^* = 2(bc - ad)\mathbf{i} + (a^2 - b^2 + c^2 - d^2)\mathbf{j} + 2(cd + ab)\mathbf{k},$$
$$q\mathbf{k}q^* = 2(bd + ac)\mathbf{i} + 2(cd - ab)\mathbf{j} + (a^2 - b^2 - c^2 + d^2)\mathbf{k}.$$

The sandwich operator is linear and the image of a vectorial quaternion is a vectorial quaternion again. Furthermore, the scalar product of two vectors $x, y \in \mathbb{R}^3$ is invariant under the action of the sandwich operator and it is orientation preserving. When collecting the images of the basis vectors e_1, e_2, e_3 in a matrix A, we get

$$A = \begin{pmatrix} a^2 + b^2 - c^2 - d^2 & 2(bc - ad) & 2(bd + ac) \\ 2(bc + ad) & a^2 - b^2 + c^2 - d^2 & 2(cd - ab) \\ 2(bd - ac) & 2(cd + ab) & a^2 - b^2 - c^2 + d^2 \end{pmatrix}. \quad (1.4)$$

Matrix (1.4) is the well-known form of a rotation matrix. The components of a unit quaternion are the *Euler parameters* of a rotation.

Dual Numbers Like complex numbers dual numbers are an extension of the real numbers. A dual number has the form $z_\epsilon = a + \epsilon b$, where a and b are real numbers and ϵ is the *dual unit* that squares to zero, $\epsilon^2 = 0$. Addition is defined component-wise. For two dual numbers the product is defined by

$$(a_1 + \epsilon b_1)(a_2 + \epsilon b_2) = a_1 a_2 + \epsilon(a_1 b_2 + a_2 b_1) + \epsilon^2(b_1 b_2)$$
$$= a_1 a_2 + \epsilon(a_1 b_2 + a_2 b_1).$$

The set of dual numbers

$$\mathbb{D} := \{a + \epsilon b \mid a, b \in \mathbb{R}, \epsilon^2 = 0\}$$

together with addition and multiplication forms a commutative ring with identity. Moreover, the dual numbers form a two-dimensional commutative unital associative algebra over the real numbers. Dual numbers with vanishing real part are zero divisors

$$(\epsilon a)(\epsilon b) = \epsilon^2(ab) = 0.$$

1.1 Description of Displacements

Definition 1.3. *For a dual number $z_\epsilon = a + \epsilon b$ the dual number $\tilde{z}_\epsilon = a - \epsilon b$ is called the conjugate dual number. The norm of a dual number that is no zero divisor then is*

$$\|z_\epsilon\| := \sqrt{z_\epsilon \tilde{z}_\epsilon} = \sqrt{(a+\epsilon b)(a-\epsilon b)} = |a|.$$

Any dual number without vanishing real part has an inverse dual number

$$z_\epsilon^{-1} = (a + \epsilon b)^{-1} := \frac{1}{a^2}\tilde{z}_\epsilon.$$

Analytic functions can be extended to dual functions with help of their formal Taylor expansion. Note that any power of ϵ that is bigger than one vanishes. Therefore, we get the Taylor expansion

$$f(a + \epsilon b) = f(a) + \epsilon b f'(a),$$

which is the dual extension of the analytic function.

Remark 1.1. *It is possible to calculate the inverse of a dual number by the Taylor expansion of z_ϵ^{-1} to make the definition of the inverse dual number clear*

$$z_\epsilon^{-1} = (a + \epsilon b)^{-1} = \frac{1}{a} - \epsilon\frac{b}{a^2} = \frac{1}{a^2}(a - \epsilon b) = \frac{1}{a^2}\tilde{z}_\epsilon.$$

Dual Vectors Later we will use dual vectors. Therefore, we introduce the n-dimensional module

$$\mathbb{D}^n := \{v_\epsilon \mid v_\epsilon = v + \epsilon\bar{v},\ \epsilon^2 = 0,\ v, \bar{v} \in \mathbb{R}^n\}.$$

A dual vector is the sum of its real- and dual part

$$v_\epsilon = v + \epsilon\bar{v},\ \text{with}\ v, \bar{v} \in \mathbb{R}^n.$$

We define a standard scalar product on this module by

$$v_\epsilon w_\epsilon^T = \langle v_\epsilon, w_\epsilon\rangle_\epsilon = \langle v, w\rangle + \epsilon(\langle v, \bar{w}\rangle + \langle \bar{v}, w\rangle),$$

where $\langle \cdot, \cdot \rangle$ denotes the standard scalar product of $\mathbb{R}^n$. Especially for the dimension $n = 3$ we are able to define a cross product by

$$v_\epsilon \times_\epsilon w_\epsilon = v \times w + \epsilon(\bar{v} \times w + v \times \bar{w}). \tag{1.5}$$

Dual Quaternions Dual quaternions were introduced by E. STUDY, see [61]. Nowadays, dual quaternions form a frequently used tool for the description of Euclidean kinematics in three dimensions, see [7, 38, 50] or [59]. In this section we give a brief introduction to dual quaternions. Our intention is to put this concept in a more general context by using Clifford algebras. Quaternions with dual number components are called dual quaternions and are denoted by

$$\mathbb{H}_d := \{a_0 + a_1\mathbf{i} + a_2\mathbf{j} + a_3\mathbf{k} + \epsilon(c_0 + c_1\mathbf{i} + c_2\mathbf{j} + c_3\mathbf{k}) \mid a_0, \ldots, a_3, c_0, \ldots, c_3 \in \mathbb{R}\}.$$

Multiplication is defined with the relations for quaternion. Furthermore, the dual unit ϵ commutes with the quaternion units $\epsilon\mathbf{i} = \mathbf{i}\epsilon$, $\epsilon\mathbf{j} = \mathbf{j}\epsilon$, $\epsilon\mathbf{k} = \mathbf{k}\epsilon$. Dual quaternions form an eight-dimensional vector space over the real numbers. The basis elements are $1, \mathbf{i}, \mathbf{j}, \mathbf{k}, \epsilon, \epsilon\mathbf{i}, \epsilon\mathbf{j}, \epsilon\mathbf{k}$.

Displacements Euclidean displacements can be described by dual unit quaternions. A dual quaternion $q_\epsilon = a_0 + a_1\mathbf{i} + a_2\mathbf{j} + a_3\mathbf{k} + \epsilon(c_0 + c_1\mathbf{i} + c_2\mathbf{j} + c_3\mathbf{k})$ is normed or a dual unit quaternion, if the norm is equal to one

$$N(q_\epsilon) := q_\epsilon q_\epsilon^* = a_0^2 + a_1^2 + a_2^2 + a_3^2 + 2\epsilon(a_0c_0 + a_1c_1 + a_2c_2 + a_3c_3) = 1,$$

where the conjugation is the quaternion conjugation

$$q_\epsilon^* = (q_1 + \epsilon q_2)^* = q_1^* + \epsilon q_2^*.$$

Therefore, a dual unit quaternion satisfies two relations in the components $a_0, \ldots, c_3$

$$a_0^2 + a_1^2 + a_2^2 + a_3^2 = 1, \qquad (1.6)$$
$$a_0c_0 + a_1c_1 + a_2c_2 + a_3c_3 = 0. \qquad (1.7)$$

Dual unit quaternions are denoted by $\mathbb{U}_d$ and form a group with respect to multiplication. More details can be found in [38]. Usually, a displacement is described by the sandwich operator. We start with a dual unit quaternion

$$q = a_0 + a_1\mathbf{i} + a_2\mathbf{j} + a_3\mathbf{k} + \epsilon(c_0 + c_1\mathbf{i} + c_2\mathbf{j} + c_3\mathbf{k})$$

1.1 Description of Displacements

and a dual quaternion of the form $p = 1 + \epsilon(x\mathbf{i} + y\mathbf{j} + z\mathbf{k})$, representing the point $P = (x, y, z)^T \in \mathbb{R}^3$. We apply the sandwich operator as

$$\begin{aligned}qpq^* = 1 &+ (xa_0^2 + 2za_0a_2 - 2ya_0a_3 + 2c_1a_0 + xa_1^2 + 2ya_1a_2 \\&+ 2za_1a_3 - 2c_0a_1 - xa_2^2 + 2c_3a_2 - xa_3^2 - 2c_2a_3)\epsilon\mathbf{i} \\&+ (ya_0^2 - 2za_0a_1 + 2xa_0a_3 + 2c_2a_0 - ya_1^2 + 2xa_1a_2 \\&- 2c_3a_1 + ya_2^2 + 2za_2a_3 - 2c_0a_2 - ya_3^2 + 2c_1a_3)\epsilon\mathbf{j} \\&+ (za_0^2 + 2ya_0a_1 - 2xa_0a_2 + 2c_3a_0 - za_1^2 + 2xa_1a_3 \\&+ 2c_2a_1 - za_2^2 + 2ya_2a_3 - 2c_1a_2 + za_3^2 - 2c_0a_3)\epsilon\mathbf{k}.\end{aligned}$$

If we rewrite the result as s product of a matrix with a vector vector in homogeneous coordinates we arrive at

$$\begin{pmatrix}1\\x\\y\\z\end{pmatrix}' = \begin{pmatrix}1 & 0 & 0 & 0\\l & a_0^2+a_1^2-a_2^2-a_3^2 & 2a_1a_2-2a_0a_3 & 2a_0a_2+2a_1a_3\\m & 2a_0a_3+2a_1a_2 & a_0^2-a_1^2+a_2^2-a_3^2 & 2a_2a_3-2a_0a_1\\n & 2a_1a_3-2a_0a_2 & 2a_0a_1+2a_2a_3 & a_0^2-a_1^2-a_2^2+a_3^2\end{pmatrix} \cdot \begin{pmatrix}1\\x\\y\\z\end{pmatrix},$$

where

$$\begin{aligned}l &= 2c_1a_0 - 2c_0a_1 + 2c_3a_2 - 2c_2a_3,\\m &= 2c_2a_0 - 2c_3a_1 - 2c_0a_2 + 2c_1a_3,\\n &= 2c_3a_0 + 2c_2a_1 - 2c_1a_2 - 2c_0a_3.\end{aligned}$$

This matrix vector product represents an Euclidean displacement, see [38] and Eq. (1.4). A parametrisation of the special Euclidean group with the help of dual unit quaternions is given by

$$\begin{aligned}Q = a_0 &+ a_1\mathbf{i} + a_2\mathbf{j} + a_3\mathbf{k} + c_0\epsilon + c_1\epsilon\mathbf{i} + c_2\epsilon\mathbf{j} + c_3\epsilon\mathbf{k}\\= \cos\frac{\varphi}{2} &- \sin\frac{\varphi}{2}l_0\mathbf{i} - \sin\frac{\varphi}{2}l_1\mathbf{j} - \sin\frac{\varphi}{2}l_2\mathbf{k} \quad (1.8)\\-\frac{v}{2}\sin\frac{\varphi}{2}\epsilon &- \left(\sin\frac{\varphi}{2}l_3 + \frac{v}{2}\cos\frac{\varphi}{2}l_0\right)\epsilon\mathbf{i}\\-\left(\sin\frac{\varphi}{2}l_4 \right. &\left. + \frac{v}{2}\cos\frac{\varphi}{2}l_1\right)\epsilon\mathbf{j} - \left(\sin\frac{\varphi}{2}l_5 + \frac{v}{2}\cos\frac{\varphi}{2}l_2\right)\epsilon\mathbf{k},\end{aligned}$$

where φ is the rotation angle and v is the magnitude of a translation in the direction defined by the Plücker coordinate vector $L = (l_0 : l_1 : l_2 : l_3 : l_4 : l_5)$, see [13]. We will define Plücker coordinates in section 1.2.1.

Matrix Representation The multiplication of two dual quaternions
$$A = a_0 + a_1\mathbf{i} + a_2\mathbf{j} + a_3\mathbf{k} + \epsilon(c_0 + c_1\mathbf{i} + c_2\mathbf{j} + c_3\mathbf{k}),$$
$$X = x_0 + x_1\mathbf{i} + x_2\mathbf{j} + x_3\mathbf{k} + \epsilon(x_4 + x_5\mathbf{i} + x_6\mathbf{j} + x_7\mathbf{k}),$$
can be written in matrix form. Therefore, we perform the multiplication as follows

$$\begin{aligned}AX = &(a_0x_0 - a_1x_1 - a_2x_2 - a_3x_3) + (a_0x_1 + a_1x_0 + a_2x_3 - a_3x_2)\mathbf{i} \\ &+ (a_0x_2 + a_2x_0 - a_1x_3 + a_3x_1)\mathbf{j} + (a_0x_3 + a_1x_2 - a_2x_1 + a_3x_0)\mathbf{k} \\ &+ (a_0x_4 - a_1x_5 - a_2x_6 - a_3x_7 + c_0x_0 - c_1x_1 - c_2x_2 - c_3x_3)\epsilon \\ &+ (a_0x_5 + a_1x_4 + a_2x_7 - a_3x_6 + c_0x_1 + c_1x_0 + c_2x_3 - c_3x_2)\epsilon\mathbf{i} \\ &+ (a_0x_6 + a_2x_4 - a_1x_7 + a_3x_5 + c_0x_2 + c_2x_0 - c_1x_3 + c_3x_1)\epsilon\mathbf{j} \\ &+ (a_0x_7 + a_1x_6 - a_2x_5 + a_3x_4 + c_0x_3 + c_1x_2 - c_2x_1 + c_3x_0)\epsilon\mathbf{k}.\end{aligned}$$

If we rewrite AX as a product of a matrix and a vector, we get
$$x' = \begin{pmatrix} A & O \\ C & A \end{pmatrix} \cdot x, \text{ with}$$

$$A = \begin{pmatrix} a_0 & -a_1 & -a_2 & -a_3 \\ a_1 & a_0 & -a_3 & a_2 \\ a_2 & a_3 & a_0 & -a_1 \\ a_3 & -a_2 & a_1 & a_0 \end{pmatrix}, C = \begin{pmatrix} c_0 & -c_1 & -c_2 & -c_3 \\ c_1 & c_0 & -c_3 & c_2 \\ c_2 & c_3 & c_0 & -c_1 \\ c_3 & -c_2 & c_1 & c_0 \end{pmatrix},$$

where $x = (x_0, \cdots, x_7)^{\mathrm{T}}$ and O is the 4×4 zero matrix. In the same way we can determine the matrix representation of the product

$$\begin{aligned}XA = &(a_0x_0 - a_1x_1 - a_2x_2 - a_3x_3) + (a_0x_1 + a_1x_0 - a_2x_3 + a_3x_2)\mathbf{i} \\ &+ (a_0x_2 + a_2x_0 + a_1x_3 - a_3x_1)\mathbf{j} + (a_0x_3 - a_1x_2 + a_2x_1 + a_3x_0)\mathbf{k} \\ &+ (a_0x_4 - a_1x_5 - a_2x_6 - a_3x_7 + c_0x_0 - c_1x_1 - c_2x_2 - c_3x_3)\epsilon \\ &+ (a_0x_5 + a_1x_4 - a_2x_7 + a_3x_6 + c_0x_1 + c_1x_0 - c_2x_3 + c_3x_2)\epsilon\mathbf{i} \\ &+ (a_0x_6 + a_2x_4 + a_1x_7 - a_3x_5 + c_0x_2 + c_2x_0 + c_1x_3 - c_3x_1)\epsilon\mathbf{j} \\ &+ (a_0x_7 - a_1x_6 + a_2x_5 + a_3x_4 + c_0x_3 - c_1x_2 + c_2x_1 + c_3x_0)\epsilon\mathbf{k}.\end{aligned}$$

The corresponding product of a matrix and a vector has the form
$$x' = \begin{pmatrix} A & O \\ C & A \end{pmatrix} \cdot x, \text{ with}$$

1.1 Description of Displacements

$$A = \begin{pmatrix} a_0 & -a_1 & -a_2 & -a_3 \\ a_1 & a_0 & a_3 & -a_2 \\ a_2 & -a_3 & a_0 & a_1 \\ a_3 & a_2 & -a_1 & a_0 \end{pmatrix}, \quad C = \begin{pmatrix} c_0 & -c_1 & -c_2 & -c_3 \\ c_1 & c_0 & c_3 & -c_2 \\ c_2 & -c_3 & c_0 & c_1 \\ c_3 & c_2 & -c_1 & c_0 \end{pmatrix}, \quad (1.9)$$

where $x = (x_0, \cdots, x_7)^T$ and O is the 4×4 zero matrix, cf. [54].

1.1.3 Dual Orthogonal Matrices

In the previous section we defined dual vectors. Naturally, we can also define matrices with dual entries. The set of matrices with n rows, m columns, and dual entries shall be denoted by $\mathbb{D}^{n \times m}$. We introduce the concept of dual orthogonal matrices that can be used to apply spatial displacements to lines with the help of a dual matrix multiplication. Therefore, we follow [57] and examine dual orthogonal 3×3 matrices. A dual orthogonal matrix $\mathcal{O} \in \mathbb{D}^{3 \times 3}$ is a matrix such that

$$\mathcal{O}\mathcal{O}^T = \mathcal{O}^T\mathcal{O} = I,$$

where I denotes the 3×3 identity matrix. The displacement of a line L can be described by a dual orthogonal matrix, see [7] or [62]. With dual orthogonal matrices we can write the displacement applied to a line L that is represented as dual unit vector ℓ_ϵ as matrix vector product

$$\ell'_\epsilon = \mathcal{O}\ell_\epsilon \text{ with } \mathcal{O} = R + \epsilon TR.$$

The real 3×3 matrix T is skew-symmetric and describes the linear mapping $x \mapsto t \times x$ and $R \in SO(3)$ is a 3×3 rotation matrix. For an orthogonal dual matrix $\mathcal{O}$ and two arbitrary dual unit vectors v_ϵ, w_ϵ we have $v'_\epsilon = \mathcal{O}v_\epsilon$, $w'_\epsilon = \mathcal{O}w_\epsilon$ and

$$\langle v'_\epsilon, w'_\epsilon \rangle_\epsilon = \langle \mathcal{O}v_\epsilon, \mathcal{O}w_\epsilon \rangle_\epsilon = v_\epsilon^T \mathcal{O}^T \mathcal{O} w_\epsilon = v_\epsilon^T w_\epsilon = \langle v_\epsilon, w_\epsilon \rangle_\epsilon.$$

Hence, the dual scalar product is invariant with respect to the action of orthogonal dual matrices. Furthermore, the group of dual orthogonal matrices is a Lie group. Its Lie algebra is the set of skew-symmetric dual matrices. In [57] the authors used this for motion interpolation purposes. Ruled surfaces are interpreted as curves on the dual unit sphere that we will introduce later on. Therefore, a generalization of the Rodriguez formula for dual skew symmetric 3×3 matrices is used:

$$\mathcal{O} = \exp w_\epsilon g = I + \sin w_\epsilon g + (\cos w_\epsilon - 1)g^2, \tag{1.10}$$

where $w_\epsilon = \theta + \epsilon d$ is a dual number called the dual angle, I the 3×3 identity matrix, and g a 3×3 skew symmetric matrix. It contains the angle θ of the rotation and the magnitude d of the translation. With Eq. (1.10) it is possible to obtain a spatial displacement that transforms one line into another. For this purpose we need the dual angle between both lines and the common normal that can be computed with Eq. (1.5). Furthermore, we need the adjoint mapping

$$ad : \mathbb{D}^3 \ni \ell_\epsilon \mapsto g \in \mathbb{D}^{3\times 3} \tag{1.11}$$

$$\ell = (g_\epsilon^1, g_\epsilon^2, g_\epsilon^3)^\mathrm{T} \mapsto \begin{pmatrix} 0 & -g_\epsilon^3 & g_\epsilon^2 \\ g_\epsilon^3 & 0 & -g_\epsilon^1 \\ -g_\epsilon^2 & g_\epsilon^1 & 0 \end{pmatrix} = g$$

that maps dual vectors to dual skew-symmetric matrices. With the help of this mapping we can parametrise the displacement that maps one line into another. Let v_ϵ, w_ϵ be two dual unit vectors and let $\langle v_\epsilon, w_\epsilon \rangle_\epsilon = \cos w_\epsilon$, $w_\epsilon = \theta + \epsilon d$ be the dual angle between them. With the dual arccosine function we can compute

$$w_\epsilon = \cos^{-1}(\theta) - \epsilon d \sin^{-1}(\theta),$$

see right before remark 1.1. With the ad mapping Eq. (1.11) and the dual Rodriguez formula (Eq. (1.10)) we write

$$\mathcal{O}(t) = \exp(w_\epsilon(t))ad(n_\epsilon) = I + \sin(w_\epsilon(t))ad(n_\epsilon) + (\cos(w_\epsilon(t)) - 1)(ad(n_\epsilon))^2,$$

where

$$n_\epsilon = \frac{v_\epsilon \times_\epsilon w_\epsilon}{\|v_\epsilon \times_\epsilon w_\epsilon\|}$$

is the normalized dual vector representation of the common normal of v_ϵ (cf. [38]) and $w_\epsilon(t) = t \cdot w_\epsilon$. The parameter t varies in the interval $t \in [0, 1]$. Now we can apply this motion to the initial line v_ϵ

$$\ell_\epsilon(t) = \mathcal{O}(t) v_\epsilon.$$

1.2 Point Models for Lines and Displacements

For $t = 0$ we obtain $\mathcal{O}(0) = I$ and the resulting line is v_ϵ itself. For $t = 1$ we obtain $\mathcal{O}(1)v_\epsilon = w_\epsilon$. The resulting ruled surfaces is a helicoid and the corresponding motion is a helical motion, see [57].

Remark 1.2. *Helicoids correspond to great circles on Study's sphere, see [57]. Therefore, helicoids can be interpreted as a special type of geodesic curves on the dual sphere, see section 1.18.*

1.2 Point Models for Lines and Displacements

First of all, a quotation from [56] is presented.

> "Working in a geometric point model enables better understanding and a simple interpretation of various objects of line space. The design of efficient algorithms involving lines is greatly simplified if it based on the right geometric model."

Sometimes it seems confusing to introduce high dimensional spaces as model spaces, but if things get easier these methods help.

1.2.1 Klein's Quadric

Lines in three-dimensional space form a four-dimensional manifold called Klein's quadric, or Plücker's quadric. It is a special Grassmann variety. In fact, every straight line in $\mathbb{P}^3(\mathbb{R})$ is mapped to a point in five-dimensional projective space $\mathbb{P}^5(\mathbb{R})$, see [7]. To make this explicit, we think of $\mathbb{R}^4$ with its standard basis as model for $\mathbb{P}^3(\mathbb{R})$. A line is spanned by two different points $X = x\mathbb{R}$ and $Y = y\mathbb{R}$ with homogeneous linearly independent coordinate vectors $x = (x_0, x_1, x_2, x_3)^\mathrm{T}$ and $y = (y_0, y_1, y_2, y_3)^\mathrm{T}$.

Definition 1.4. *For two distinct points $X = x\mathbb{R} = (x_0, x_1, x_2, x_3)^\mathrm{T}\mathbb{R}$, $Y = y\mathbb{R} = (y_0, y_1, y_2, y_3)^\mathrm{T}\mathbb{R} \in \mathbb{P}^3(\mathbb{R})$ we define the Plücker coordinates of the line spanned by X and Y as:*

$$p = (p_{01} : p_{02} : p_{03} : p_{23} : p_{31} : p_{12}), \quad \text{with } p_{ij} = \begin{vmatrix} x_i & x_j \\ y_i & y_j \end{vmatrix}.$$

The mapping $\mu : \mathcal{L}^3 \mapsto \mathbb{P}^5(\mathbb{R})$, where $\mathcal{L}^3$ is the set of lines of $\mathbb{P}^3(\mathbb{R})$, that maps each line $L \in \mathcal{L}^3$ to a point $P = (p_{01} : \ldots : p_{12})$ of $\mathbb{P}^5(\mathbb{R})$ is called the Klein mapping.

Only those points of $\mathbb{P}^5(\mathbb{R})$ correspond to a line in $\mathbb{P}^3(\mathbb{R})$ that are contained in Klein's quadric M_2^4. Its equation is derived by expanding the determinant $\det(x, y, x, y)$ by complementary 2×2 minors:

$$\begin{vmatrix} x_0 & x_1 & x_2 & x_3 \\ y_0 & y_1 & y_2 & y_3 \\ x_0 & x_1 & x_2 & x_3 \\ y_0 & y_1 & y_2 & y_3 \end{vmatrix} = \begin{vmatrix} x_0 & x_1 \\ y_0 & y_1 \end{vmatrix} \begin{vmatrix} x_2 & x_3 \\ y_2 & y_3 \end{vmatrix} + \begin{vmatrix} x_0 & x_2 \\ y_0 & y_2 \end{vmatrix} \begin{vmatrix} x_3 & x_1 \\ y_3 & y_1 \end{vmatrix} + \begin{vmatrix} x_0 & x_3 \\ y_0 & y_3 \end{vmatrix} \begin{vmatrix} x_1 & x_2 \\ y_1 & y_2 \end{vmatrix} = 0,$$

which yields

$$M_2^4 : p_{01}p_{23} + p_{02}p_{31} + p_{03}p_{12} = 0. \tag{1.12}$$

Eq. (1.12) can be reformulated as

$$x^T Q x = 0, \text{ with } Q = \begin{pmatrix} O & \frac{1}{2}I \\ \frac{1}{2}I & O \end{pmatrix}, x = (x_0, x_1, x_2, x_3, x_4, x_5)^T, \tag{1.13}$$

where O is the 3×3 zero matrix and I is 3×3 identity matrix. We have a one-to-one correspondence between lines in three-dimensional projective space and points on this quadric. The polarity of Klein's quadric ν can be expressed in matrix form by Q. When working in five-dimensional projective space we use ν and postfix notation. Moreover, the polarity of Klein's quadric ν is regular, since

$$\det Q = -\frac{1}{64} \neq 0.$$

The bilinear form induced by this polarity is denoted by Ω and defined by

$$\Omega(x, x) := x^T Q x.$$

Two lines L_1 and L_2 given in Plücker coordinates $l_1\mathbb{R}$ and $l_2\mathbb{R}$ intersect if, and only if, their images under the Klein map satisfy $\Omega(L_1, L_2) := l_1^T Q l_2 = 0$. This means, they have to be conjugate with respect to Klein's quadric.

1.2 Point Models for Lines and Displacements

Direction and Moment We write Plücker coordinates in the form (l, m). If we interpret l and m as vectors in $\mathbb{R}^3$ they are orthogonal with respect to the standard scalar product and satisfy the Plücker condition (1.12). From Def. 1.4 we see that $m = x \times y$ with $x = (x_1, x_2, x_3)^\mathrm{T}$ and $y = (y_1, y_2, y_3)^\mathrm{T}$. Furthermore, l is the direction of the line. This follows by setting the homogeneous factors x_0 and y_0 to 1. In this case the first three Plücker coordinates are $l = x - y$.
Note that it is possible to calculate Plücker coordinates from a given direction vector $l \in \mathbb{R}^3$ and an arbitrary point p on the line by

$$(l, m) = (l, p \times l).$$

This is independent of the choice of p, since if $\bar{p} = p + \lambda l$, $\lambda \in \mathbb{R}$ is another point on the line the cross product simplifies to

$$(p + \lambda l) \times l = p \times l + \lambda l \times l = p \times l.$$

The vector $l \in \mathbb{R}^3$ is called the *direction vector* and $m \in \mathbb{R}^3$ the *momentum vector*. Moreover, we know from projective geometry that the ideal point of a line, if not the line itself is ideal, is given by $(0, l)^\mathrm{T} \mathbb{R}$.

Subspaces contained in the Quadric An algebraic variety of dimension d and degree k is denoted by a capital letter with subscript k and superscript d, say V_k^d. Thus, Klein's quadric M_2^4 is an example for an algebraic variety. We are interested in the structure of Klein's quadric. The quadric is hyperbolic with two-dimensional generator spaces. Therefore, we take a closer look to one- and two-dimensional subspaces contained in the quadric. In this paragraph we follow [56].

One-dimensional subspaces of M_2^4 To classify one-dimensional subspaces we give a lemma from [56, section 2.1.3] without proof. Nevertheless, it is easy to verify that:

Lemma 1.1. *The Klein mapping takes a pencil of lines to a straight line contained in M_2^4. Vice versa, two points $X = x\mathbb{R}$ and $Y = y\mathbb{R}$ of M_2^4 correspond to intersecting lines in $\mathbb{P}^3(\mathbb{R})$ if, and only if, their span is contained in M_2^4.*

Two-dimensional subspaces of M_2^4 There are two types of two-spaces that are completely contained by M_2^4. The first one is given by the image of a field of lines, *i.e.*, all lines contained by the same plane in $\mathbb{P}^3(\mathbb{R})$. The second type is given by all lines concurrent to the same point, *i.e.*, a bundle of lines. Now we ask for the intersection of two two-spaces contained in Klein's quadric. Two-spaces of the same type always intersect in one point. This means two different fields of lines or two different bundles of lines contain one common line. Two two-spaces of different type may have either empty intersection or they intersect in a line. Thus, a field of lines and a bundle of lines may have empty intersection or they intersect in a common pencil of lines. The type of the two-space, *i.e.*, if it corresponds to a bundle of lines or a field of lines can be determined by its intersection with the Klein image of the field of ideal lines denoted by P_ω. Note that P_ω corresponds to a field of lines, and therefore, to a two-space entirely contained in Klein's quadric. If a two-space corresponds to a bundle of lines it has either no point or a whole line in common with P_ω. A two-space corresponding to a field of lines has one point in common with P_ω. Fig. 1.1 shows the correspondence between lines and two-spaces on Klein's quadric and their pre-images under the Klein mapping. This figure is inspired by a figure from [56, p. 142].

Subspace Intersections of M_2^4 With Klein's mapping sets of lines can be studied as sets of points in five-dimensional projective space. Clearly, all points on the quadric are self-conjugate with respect to the polarity ν. It is natural to ask for the intersections of Klein's quadric with subspaces. In Klein's model k-space intersections, $k \leq 4$ with $M_2^4 \subset \mathbb{P}^5(\mathbb{R})$ define special sets of lines in $\mathbb{P}^3(\mathbb{R})$. These sets of lines correspond to so-called linear line manifolds, cf. [56].

Conics on Klein's quadric   Two-spaces which are neither tangent to nor belong to Klein's quadric intersect in a conic section. Tangent two-space intersections with M_2^4 result in degenerate conics, *i.e.*, two intersecting lines on Klein's quadric that correspond to two pencils of lines with one line in common in projective space $\mathbb{P}^3(\mathbb{R})$. In this paragraph we discus the non-degenerate case. Therefore, we choose three skew lines respectively their Plücker coordinates L_1, L_2 and L_3

1.2 Point Models for Lines and Displacements

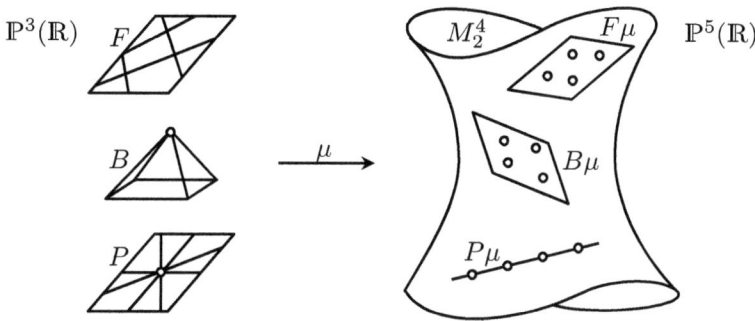

Figure 1.1: Subspaces on Klein's quadric and their geometric interpretation in $\mathbb{P}^3(\mathbb{R})$. A pencil of lines P is mapped to the line $P\mu$ on M_2^4. A bundle of lines B and a field of lines F are mapped to a two-spaces $B\mu$ and $F\mu$ that are contained entirely in M_2^4.

with $\Omega(L_i, L_j) \neq 0$, $i, j = 1, 2, 3$, $i \neq j$. These three lines possess image points spanning a two-space

$$P_1^2(\alpha : \beta : \gamma) = \alpha L_1 + \beta L_2 + \gamma L_3, \ \alpha, \beta, \gamma \in \mathbb{R}.$$

The set of all points conjugate to this two-space is spanned by $P_1^2 \nu = L_1\nu \cap L_2\nu \cap L_3\nu$. This defines the polar two-space of P_1^2. Since this polar two-space is the intersection of three tangent hyperplanes we conclude that the intersection $P_1^2\nu \cap M_2^4$ corresponds to the set of all lines in $\mathcal{L}^3$ intersecting L_1, L_2, and L_3. Now we can do this construction for three points of the resulting conic again and get the same statement for the original points L_1, L_2, and L_3. Furthermore, we derive that these two sets of lines are different sets of generators of the same ruled surface. We define:

Definition 1.5. *The set of all lines intersecting three given mutually skew lines $L_1, L_2, L_3 \in \mathcal{L}^3$ is called a* regulus. *A regulus is part of a ruled quadric. The image of a regulus under the polarity ν that defines the second family of generators of a ruled quadric is called the* opposite regulus.

Therefore, every regular conic on Klein's quadric corresponds to a regulus. Reguli can be distinguished in the Klein model by their intersection with the two-space corresponding to the field of ideal lines. In affine space a regulus that carries an ideal line is a hyperbolic paraboloid. Otherwise if there is no ideal line the regulus corresponds to a hyperboloid of one sheet.

Linear Line Congruences The two-parametric set of lines corresponding to a three-space intersection of Klein's quadric is called a linear line congruence. In line geometry we distinguish between hyperbolic, parabolic, and elliptic linear line congruences, see [56].

Linear Line Complexes Linear line complexes correspond to hyperplanar intersections of M_2^4. Therefore, they define three-dimensional line manifolds.

Definition 1.6. *A linear line complex is defined by a linear equation in Plücker coordinates:*

$$\Omega(C, L) = \bar{c} \cdot l + c \cdot \bar{l} = 0,$$

with $L = (l, \bar{l})^\mathrm{T} \mathbb{R}$ and $C = (c, \bar{c})^\mathrm{T} \mathbb{R}$, where C is given.

A linear line complex is called *singular* if the hyperplane is tangent to M_2^4 else *regular*. Note that the hyperplane defined by $\mathbb{R}(\bar{c}, c)^\mathrm{T}$ is polar to the point $C = (c, \bar{c})^\mathrm{T} \mathbb{R}$. This point is contained in M_2^4 in the singular case.

Remark 1.3. *We can derive any regular (singular) linear line complex from one single regular (singular) linear line complex by a projective transformation. With respect to affine geometry one has to distinguish between two equivalence classes of regular linear line complexes, right winded and left winded complexes. In a projectively enclosed Euclidean space each linear complex is characterised by an Euclidean invariant, the so-called* pitch $p \in \mathbb{R}$ *(see [56], [66]) and all proper lines of a regular linear line complex can be interpreted as the path normals of a helical motion. If $(c, \bar{c})^\mathrm{T}\mathbb{R} = ((c_1, c_2, c_3), (c_4, c_5, c_6))^\mathrm{T}\mathbb{R}$ is the Plücker coordinate vector of a linear line complex C, then we get its pitch p by*

$$p = \frac{c \cdot \bar{c}}{c^2}. \tag{1.14}$$

The axis of the helicoidal motion corresponding to the linear line complex C is determined by

$$A = (a, \bar{a})^{\mathrm{T}} \mathbb{R} = ((a_1, a_2, a_3), (a_4, a_5, a_6))^{\mathrm{T}} \mathbb{R} = (c, \bar{c} - pc)^{\mathrm{T}} \mathbb{R}.$$

In case of a singular linear line complex the pitch equals 0, $A = C$, and the complex contains all lines that are path normals of a one-parameter pure rotation or pure translation group of three-dimensional Euclidean space $\mathbb{E}^3$.

The condition $\Omega(C, L) = 0$ is exactly the condition for the intersection of lines in this case. With this knowledge we can identify the points of $\mathbb{P}^5(\mathbb{R})$ that are not contained by M_2^4 with regular linear line complexes. The points on M_2^4 correspond to singular linear line complexes. This is called the *extended Klein mapping* $\hat{\mu}$, see [66].

1.2.2 Study's Quadric

The six-dimensional group of Euclidean displacements in three-dimensional space can be mapped to a special hyperquadric S_2^6 in $\mathbb{P}^7(\mathbb{R})$ which is usually referred to as Study's quadric. Each displacement is represented by a point on this quadric, but not every point on S_2^6 corresponds to a displacement. There exists a three-dimensional generator space $V^3 \subset S_2^6$ whose points do not correspond to displacements. If we denote a point $P \in \mathbb{P}^7(\mathbb{R})$ by $P = (a_0, \ldots, a_3, c_0, \ldots, c_3)^{\mathrm{T}} \mathbb{R}$, this exceptional space V^3 is given by $a_0 = a_1 = a_2 = a_3 = 0$. Therefore, the point set $S_2^6 \backslash V^3$ is the image of the Euclidean displacements. Thus, the image space is a sliced quadric $S_2^6 \backslash V^3$, a pseudo algebraic variety, so to say. Hence, we have a bijective mapping

$$\zeta : \mathrm{SE}(E) \to S_2^6 \backslash V^3 \subseteq \mathbb{P}^7(\mathbb{R}),$$
$$\mathrm{SE}(3) \ni \alpha \mapsto A = (a_0, a_1, a_2, a_3, c_0, c_1, c_2, c_3)^{\mathrm{T}} \mathbb{R}.$$

STUDY himself [61] gave a method to compute the *Study coordinates*. Let

$$\mathrm{M} = \begin{pmatrix} m_{00} & 0 & 0 & 0 \\ m_{10} & m_{11} & m_{12} & m_{13} \\ m_{20} & m_{21} & m_{22} & m_{23} \\ m_{30} & m_{31} & m_{32} & m_{33} \end{pmatrix}$$

be a homogeneous transformation matrix of an Euclidean displacement, see Eq. (1.1). The coordinates of a point on $S_2^6 \setminus V^3$ corresponding to this displacement can now be determined by the relations:

$$a_0 : a_1 : a_2 : a_3 \qquad (1.15)$$
$$= m_{00}+m_{11}+m_{22}+m_{33} : m_{23}-m_{32} : m_{31}-m_{13} : m_{12}-m_{21},$$
$$= m_{23}-m_{32} : m_{00}+m_{11}-m_{22}-m_{33} : m_{12}+m_{21} : m_{31}+m_{13},$$
$$= m_{31}-m_{13} : m_{12}+m_{21} : m_{00}-m_{11}+m_{22}-m_{33} : m_{23}+m_{31},$$
$$= m_{12}-m_{21} : m_{31}+m_{13} : m_{23}+m_{32} : m_{00}-m_{11}-m_{22}+m_{33},$$

and

$$2m_{00}c_0 = a_1 m_{10} + a_2 m_{20} + a_3 m_{30}, \qquad (1.16)$$
$$2m_{00}c_1 = -a_0 m_{10} + a_3 m_{20} - a_2 m_{30},$$
$$2m_{00}c_2 = -a_3 m_{10} - a_0 m_{20} + a_1 m_{30},$$
$$2m_{00}c_3 = a_2 m_{10} - a_1 m_{20} - a_0 m_{30}.$$

These relations can be found in [23] or [61]. Note that each equation of Eqs. (1.15) is important because the ratio could get $(0:0:0:0)$ in one equation, but not in all equations, what is not allowed since the point would lie in the exceptional space V^3. The entries of the matrix representing the displacement can be calculated by

$$\begin{aligned}
m_{00} &= a_0^2 + a_1^2 + a_2^2 + a_3^2, & m_{10} &= 2(a_2 c_3 - a_3 c_2 - a_0 c_1 + a_1 c_0), \quad (1.17)\\
m_{11} &= a_0^2 + a_1^2 - a_2^2 - a_3^2, & m_{12} &= 2(a_1 a_2 + a_0 a_3),\\
m_{13} &= 2(a_3 a_1 - a_0 a_2), & m_{20} &= 2(a_3 c_1 - a_1 c_3 - a_0 c_2 + a_2 c_0),\\
m_{21} &= 2(a_1 a_2 - a_0 a_3), & m_{22} &= a_0^2 - a_1^2 + a_2^2 - a_3^2,\\
m_{23} &= 2(a_2 a_3 + a_0 a_1), & m_{30} &= 2(a_1 c_2 - a_2 c_1 - a_0 c_3 + a_3 c_0),\\
m_{31} &= 2(a_3 a_1 + a_0 a_2), & m_{32} &= 2(a_2 a_3 - a_0 a_1),\\
m_{33} &= a_0^2 - a_1^2 - a_2^2 + a_3^2.
\end{aligned}$$

The coordinates of all points on S_2^6 satisfy the equation

$$S_2^6 : a_0 c_0 + a_1 c_1 + a_2 c_2 + a_3 c_3 = 0.$$

This defines a " hyperbolic " quadric of maximum index three in $\mathbb{P}^7(\mathbb{R})$ that carries two families of three-dimensional generator spaces each being a projective three-dimensional space.

1.2 Point Models for Lines and Displacements

Subspaces contained in S_2^6 A discussion of linear subspaces contained entirely in Study's quadric can be found in [4]. The maximal possible dimension of subspaces contained in a six-dimensional quadric equals three. In fact S_2^6 carries two three-parameter families of three-dimensional subspaces. To classify the subspaces that are entirely contained in S_2^6 we follow [24]. First we observe that a point $P \in S_2^6$ corresponds to the identity, a rotation, or a translation if, and only if, $c_0 = 0$, see [25]. This follows directly with Eq. (1.8). Thus, all rotations and translations are contained in the hyperplane $x_4 = 0$, that is the tangent hyperplane at the point $P_I = (1, 0, \ldots, 0)^T \mathbb{R}$, i.e., the point on S_2^6 corresponding to the image $\zeta(e)$ of the identity element $e \in SE(3)$ under the Study mapping. All points that are conjugate to P_I are contained in a quadratic cone with equation

$$C_0 : x_1 x_5 + x_2 x_6 + x_3 x_7 = 0.$$

This equation can also be interpreted as the equation of $M_2^4 \subseteq \mathbb{P}^5(\mathbb{R})$ after changing the indices, see [59, Chapter 11].

Lines within S_2^6 Let A_0, A_1 be two different points on $S_2^6 \backslash V^3$. It is possible to find a collineation κ of the form (1.9) that maps A_0 to the identity P_I and A_1 to some point on the cone C_0. Since the mapping κ preserves the exceptional generator there are two possibilities. The image $P_1^1 \kappa$ of the line $P_1^1 = [A_0, A_1]$ under κ has one or no point in common with V^3. With Eq. (1.8) we see that if the line has a point in common with V^3 it corresponds to a one-parameter group of translations, otherwise it corresponds to a one-parameter group of rotations. Thus, a general line on Study's quadric corresponds to the composition of a fixed displacement κ^{-1} with a one-parameter rotation or translation group.

Two-Spaces within S_2^6 We can apply the same procedure that we used for lines on the quadric. Following [24] we define

Definition 1.7. *(a) Let P_1^2 be a plane in $\mathbb{P}^3(\mathbb{R})$ and $O \in P_1^2$ be a point. We call the two-parametric set of rotations with axes corresponding to the pencil of lines P_1^2 with vertex O a* pencil of rotations. *If P_1^2 is a proper plane we call it a* pencil of rotations with proper or

improper vertex *depending on whether O is proper or improper.
If P_1^2 is an improper plane, we speak of a* pencil of rotations with
improper axes, *i.e., a pencil of translations.*

(b) *The three-parametric set of all rotations with axes in a fixed
plane $P_1^2 \in \mathbb{P}^3(\mathbb{R})$ is called a* field of rotations *and P_1^2 is called its*
supporting plane.

(c) *The three-parametric set of rotations with axes through a fixed
point $O \in \mathbb{P}^3(\mathbb{R})$ is called a* bundle of rotations *and O is called its*
vertex, *i.e., a coset of* SO(3).

For every projective two-space spanned by three points $P_1^2 = [A_0, A_1, A_2]$ it is possible to find a collineation κ of the form (1.9) with $A_0\kappa = P_I$. This results in three different cases for the intersection of $P_1^2\kappa$ with V^3. If the intersection is empty $P_1^2\kappa$ corresponds to a pencil of rotations. If the intersection is a single point $P_1^2\kappa$ corresponds to a pencil of rotations with improper vertex, and if the intersection is a whole line $P_1^2\kappa$ corresponds to a pencil of rotations with improper axes.

Three-Spaces within S_2^6 There are two different types of three-spaces that are contained entirely in Study's quadric. They are distinguished by their intersection with the exceptional generator V^3. Generators that do not intersect V^3 correspond to cosets of SO(3), *i.e.*, the composition of a fixed displacement with SO(3) respectively to bundles of rotations. This follows with a collineation κ of type (1.9) that maps one point from the three-space to P_I. If the intersection of the three-space with the exceptional generator is a single point, the corresponding displacements are contained by the composition of a fixed displacement with all displacements of a field of rotations with axes in a proper plane. If the intersection of the three-spaces with V^3 is a line, the set of displacements is contained in the composition of a fixed displacement with all displacements of a bundle of rotations with improper vertex, *i.e.*, a coset of SE(2). The last case that may occur is if the intersection of the generator with the exceptional generator is a two-space. In this case the corresponding set of displacements is the composition of a fixed displacement with all displacements of a field of rotations with improper axes, *i.e.*, the group of all translations. For proofs and a complete discussion we refer to [24].

1.2.3 Study's Sphere

Another point model for lines or for Euclidean displacements is Study's sphere, see [29, 56, 65]. Study's sphere is a point model for spears, *i.e.*, oriented lines. Each line in Euclidean space carries two different spears. It can be oriented in two different ways. We consider a dual vector $v_\epsilon = v + \epsilon \bar{v} \in \mathbb{D}^3$. The canonical scalar product of this dual vector with itself results in

$$\langle v_\epsilon, v_\epsilon \rangle_\epsilon = \langle v, v \rangle + 2\epsilon \langle v, \bar{v} \rangle. \tag{1.18}$$

If the dual vector is built from Plücker coordinates $(v, \bar{v})$ of a line in three-dimensional projective space $\mathbb{P}^3(\mathbb{R})$, the real part contains the direction and the dual part the momentum of a line. Then, the dual part of Eq. (1.18) vanishes. Thus, the value of the scalar product is a real number. Furthermore, we assume that the direction vector of the line is normalized. In this case the scalar product (1.18) is equal to one and we call a dual vector v_ϵ with $\langle v_\epsilon, v_\epsilon \rangle_\epsilon = 1$, a *normalized dual vector* or *dual unit vector*. Therefore, we can identify the set of oriented lines in $\mathbb{R}^3$ with points of the dual unit sphere

$$S_\mathbb{D}^2 := \{ x_\epsilon \in \mathbb{D}^3 \mid \langle x_\epsilon, x_\epsilon \rangle_\epsilon = 1 \}.$$

Definition 1.8. *The mapping that maps an oriented line L of Euclidean three-space to the dual vector $v_\epsilon = v + \epsilon \bar{v}$, where $(v, \bar{v})$ are the normalized Plücker coordinates of L, is called the* Study mapping. *Its image space is a model of the set of spears of the three-dimensional Euclidean space and is called Study's sphere.*

Remark 1.4. *For dual unit quaternions the same construction can be applied. In this case we get the set of all dual unit vectors of $\mathbb{D}^4$. Note, that the direction of the Euler coordinates defines the direction of the displacement. We have to differ between two displacements corresponding to different oriented Euler coordinates, and therefore, we consider* oriented *displacements.*

1.3 Geometric Algebras, Clifford Algebras

Geometric algebras are special Clifford algebras over the field of real numbers. A general introduction is given in [22]. Clifford algebras are associative algebras that generalize complex numbers, Hamilton's quaternions, and biquaternions. Geometric algebras, abbreviated by GA, find applications in computer graphics [19] robotics [28, 59, 50], physics [36], and a lot of other disciplines. Here, we focus on their application in kinematics of Cayley-Klein geometries and on conformal geometric algebras, and its generalization. The great advantage of GA is that geometric entities such as points, lines, and planes can be described as elements of the algebra. Furthermore, transformations can also be described as special elements of the algebra and the action of the transformations applied to a geometric entity is realized by an algebra operation. A complete treatise of this topic is off the scope of this work, but we try to give enough references for the interested reader.

1.3.1 Definition of a Geometric Algebra

We start with a finite-dimensional real vector space $V = \mathbb{R}^n$ equipped with a quadratic form $\rho: V \to \mathbb{R}$. The pair (V, ρ) is called a *quadratic space*. The matrix of the quadratic form ρ is denoted by $(Q_{ij})_{i,j}$ with $1 \leq j, j \leq n$. Therefore, $\rho(x_i, x_j) = Q_{i,j}$ for some basis vectors x_i and x_j and we denote the quadratic form by its matrix representation Q. The algebra is defined by the relations

$$x_i x_j + x_j x_i = 2Q_{ij}, \quad 1 \leq i, j \leq n. \tag{1.19}$$

Usually, the corresponding Clifford algebra is denoted by $\mathcal{Cl}(V, Q)$. We shall use another notation. According to Sylvester's law of inertia, we can always find a basis $\{e_1, \ldots, e_n\}$ of V such that $e_i^2 \in \{1, -1, 0\}$. This basis is called the *standard basis* of the Clifford algebra. The number of basis vectors that square to $(1, -1, 0)$ is denoted by the *signature* (p, q, r). Therefore, we use the notation $\mathcal{Cl}_{(p,q,r)}$ instead of $\mathcal{Cl}(V, \rho)$. If $r \neq 0$ we call the geometric algebra *degenerate*. An n-dimensional real vector space equipped with a non-degenerate quadratic form of

1.3 Geometric Algebras, Clifford Algebras

signature $(p,q,0)$, $n = p+q$ is denoted by $\mathbb{R}^{(p,q)}$. Furthermore, the relations (1.19) become

$$e_i e_j + e_j e_i = 0, \quad i \neq j, \tag{1.20}$$

if we use Sylvester's law of inertia. For details we refer to [59]. Note that Eq. (1.20) also shows that two basis elements anti-commute, i.e.,

$$e_i e_j = -e_j e_i, \quad i \neq j. \tag{1.21}$$

In the remainder of this thesis we abbreviate the product of basis elements with lists and call them generators

$$e_{12\ldots k} := e_1 e_2 \ldots e_k, \text{ with } 0 \leq k \leq n.$$

In order to bring light into this definition, we give an example.

Example 1.1. *We present the quaternions as elements of a Clifford algebra. Therefore, we construct the Clifford algebra $\mathcal{C}\ell_{(0,2,0)}$. This means we have two basis vectors, that square to -1. The field $\mathbb{R}$ is embedded with the basis element e_0 and a element of $\mathcal{C}\ell_{(0,2,0)}$ can be written as:*

$$a_0 e_0 + a_1 e_1 + a_2 e_2 + a_{12} e_{12}.$$

The basis $\{e_0, e_1, e_2, e_{12}\}$ for $\mathcal{C}\ell_{(0,2,0)}$ is called a standard basis *for this geometric algebra. As we already know from the definition $e_1^2 = e_2^2 = -1$. The square of e_{12} evaluates to*

$$e_{12}^2 = e_{12} e_{12} = -e_{12} e_{21} = e_{11} = -1.$$

If we now identify e_1 with $\mathbf{i}$, e_2 with $\mathbf{j}$, and e_{12} with $\mathbf{k}$, we have an isomorphism between $\mathcal{C}\ell_{(0,2,0)}$ and $\mathbb{H}$. Of course the multiplication rules for the quaternion units $\mathbf{i}, \mathbf{j}$ and $\mathbf{k}$ have to be verified. This is left to the reader as an exercise.

Definition 1.9. *For every Clifford algebra over an n-dimensional vector space a general element is the sum of scalars, bivectors, trivectors up to pseudoscalars, it is called a* multivector *and reads*

$$a_0 e_0 + a_1 e_1 + a_n e_n + a_{12} e_{12} + \ldots + a_{n(n-1)} e_{n(n-1)} + \ldots + a_{1\ldots n} e_{1\ldots n}.$$

1.3.2 Properties of Clifford Algebras

A Clifford algebra is a graded algebra Every basis element $e_{\alpha\beta\ldots\gamma}$ can be transformed to a basis element of the form $e_{ij\ldots k}$, where $i < j < \ldots < k$. Each swap of two elements causes a multiplication with -1. If we arrive at e_{ii} we can insert 1, or -1, or 0 as prescribed by the signature (p, q, r). The 2^n monomials

$$e_{i_1 i_2 \ldots i_k}, \quad 0 \le k \le n$$

form the standard basis of the Clifford algebra.

Definition 1.10. *An algebra element that is the product of invertible grade-1 elements is called a* versor.

The Clifford algebra $\mathcal{C}\ell_{(p,q,r)}$ is the direct sum $\bigoplus_{i=0}^{n} \bigwedge^i V$ of all exterior products $\bigwedge^i V$ of V of any grade $0 \le i \le n$ where $e_{k_1\ldots k_i}$, $k_1 < \ldots < k_i$ form a basis of $\bigwedge^i V$. Especially $\bigwedge^0 V$ is the scalar part $\mathbb{R}$ and $\bigwedge^1 V$ is isomorphic to the vector space V. Elements from $\bigwedge^2 V$ are called bivectors and elements from $\bigwedge^n V$ are called *pseudoscalars*. The dimension of each subspace $\bigwedge^i V$ is $\binom{n}{i}$. Thus, the dimension of the Clifford algebra is $\sum_{i=0}^{n}(\dim \bigwedge^i V) = \sum_{i=0}^{n} \binom{n}{i} = 2^n$. A Clifford algebra is called *universal* if $\dim \mathcal{C}\ell_{(p,q,r)} = 2^n$, $n = p + q + r$. In [22, p. 89] it is shown that there always exists an universal Clifford algebra for a given quadratic space (V, ρ). We restrict ourselves to the standard basis. Furthermore, the Clifford algebra $\mathcal{C}\ell_{(p,q,r)}$ possesses a $\mathbb{Z}_2$-grading, i.e., it is the direct sum of an even and an odd part

$$\mathcal{C}\ell_{(p,q,r)} = \mathcal{C}\ell^+_{(p,q,r)} \oplus \mathcal{C}\ell^-_{(p,q,r)} := \bigoplus_{\substack{i=0 \\ i \text{ even}}}^{n} \bigwedge^i V \oplus \bigoplus_{\substack{i=0 \\ i \text{ odd}}}^{n} \bigwedge^i V. \qquad (1.22)$$

Note that the even part $\mathcal{C}\ell^+_{(p,q,r)}$ is always a subalgebra, because the product of two even-graded monomials must be even-graded and the generators cancel only in pairs. The dimension of the even subalgebra is 2^{n-1}. Furthermore, we have the isomorphism

$$\mathcal{C}\ell_{(p,q,r)} \cong \mathcal{C}\ell^+_{(p,q+1,r)}. \qquad (1.23)$$

For details see [55]. To make this isomorphism more clear we give another example.

1.3 Geometric Algebras, Clifford Algebras

Example 1.2. *As we already know from Ex. 1.1 the Clifford algebra $C\ell_{(0,2,0)}$ is isomorphic to $\mathbb{H}$. The even part of this algebra is generated by $\{e_0, e_{12}\}$, and therefore, isomorphic to $\mathbb{C}$. Furthermore, the Clifford algebra $C\ell_{(0,1,0)}$ is the algebra with one generator squaring to -1. Therefore, $C\ell_{(0,1,0)} \cong \mathbb{C}$. All in all we have $C\ell_{(0,2,0)}^+ \cong C\ell_{(0,1,0)}$.*

Definition 1.11. *The center of a ring $\mathcal{R}$ is the set of all elements that commute with all other elements*

$$\mathcal{C}(\mathcal{R}) := \{c \in \mathcal{R} \mid cx = xc \text{ for all } a \in \mathcal{R}\}.$$

We are interested in the center of a Clifford algebra, see [22, p.95]. The center of a Clifford algebra $C\ell_{(p,q,r)}$ is

(1) $\mathcal{C}(C\ell_{(p,q,r)}) = \{\alpha e_0 + \beta e_{12\ldots n} \mid \alpha, \beta \in \mathbb{R}\}$ if n is odd,

(2) $\mathcal{C}(C\ell_{(p,q,r)}) = \{\alpha e_0 \mid \alpha \in \mathbb{R}\}$ if n is even.

For the even part the center is

(3) $\mathcal{C}(C\ell_{(p,q,r)}^+) = \{\alpha e_0 \mid \alpha \in \mathbb{R}\}$ if n is odd,

(4) $\mathcal{C}(C\ell_{(p,q,r)}^+) = \{\alpha e_0 + \beta e_{12\ldots n} \mid \alpha, \beta \in \mathbb{R}\}$ if n is even.

Clifford Algebra Involution

The Conjugation Every Clifford algebra possesses an *anti-automorphism*, i.e., an automorphism of the algebra that reverses the order of factors in a product. We follow [59] and denote the conjugation by an asterisk. The action on any generator is given by $e_i^* = -e_i$. On scalars it has no effect. If we extend the conjugation by using linearity to arbitrary algebra elements, we get the formula

$$(e_{i_1} e_{i_2} \ldots e_{i_k})^* = (-1)^k e_{i_k} \ldots e_{i_2} e_{i_1}, \quad 0 \leq i_1 < i_2 < \ldots < i_k \leq n. \quad (1.24)$$

Example 1.3. *We apply the conjugation to an algebra element of the Clifford algebra $C\ell_{(0,2,0)} \cong \mathbb{H}$. The standard basis is given by $\{e_0, e_1, e_2, e_{12}\}$. The conjugation for the two grade-1 basis elements e_1 and e_2 is given via definition by $e_1^* = -e_1$ and $e_2^* = -e_2$. The conjugation has no effect on scalars, so $e_0^* = e_0$. It remains to extend* these observations to the grade-2 basis element e_{12}. We use the anti-involution property and the anti-commutativity

$$e_{12}^* = (e_1 e_2)^* = e_2^* e_1^* = (-1)^2 e_2 e_1 = e_{21} = -e_{12}.$$

Hence, the conjugation of quaternions fits into the concept.

A general grade-1 element in $\mathcal{C}\ell_{(p,q,r)}$ has the form

$$x = x_1 e_2 + x_2 e_2 + \ldots + x_n e_n, \quad n = p + q + r.$$

The product with its conjugate element is given by:

$$xx^* = -x_1^2 - x_2^2 - \ldots - x_p^2 + x_{p+1}^2 + \ldots + x_{p+q}^2 = -\rho(x,x). \quad (1.25)$$

If we identify V with $\bigwedge^1 \mathcal{C}\ell_{(p,q,r)}$ Eq. (1.25) gives the negative of the quadratic form $\rho(x,x)$. Furthermore, the square of $x \in \bigwedge^1 \mathcal{C}\ell_{(p,q,r)}$ results in the scalar product of the vector with itself

$$xx = x_1^2 + x_2^2 + \ldots + x_p^2 - x_{p+1}^2 - \ldots - x_{p+q}^2 = \rho(x,x).$$

The Main Involution   Another involution of a Clifford algebra is the *main involution*. It is denoted by α and defined by

$$\alpha(e_{i_1} e_{i_2} \ldots e_{i_k}) = (-1)^k e_{i_1} e_{i_2} \ldots e_{i_k}, \quad 0 \leq i_1 < i_2 < \ldots < i_k \leq n. \quad (1.26)$$

The main involution has no effect on the even subalgebra and it commutes with the conjugation. This means for an arbitrary algebra element x, the equation $\alpha(x^*) = \alpha(x)^*$ holds.

Clifford Algebra Products   We denote grade-1 elements, *i.e.*, vectors, by small gothic letters. Elements of higher grade are denoted by big gothic letters.

The inner Product   The scalar product of two vectors, *i.e.*, grade-1 elements can be written in terms of the geometric product

$$\mathfrak{a} \cdot \mathfrak{b} := \frac{1}{2}(\mathfrak{a}\mathfrak{b} + \mathfrak{b}\mathfrak{a}). \quad (1.27)$$

1.3 Geometric Algebras, Clifford Algebras

A generalization of the inner product to blades, see Def. 1.12, can be found in [36]. For $\mathfrak{A} \in \bigwedge^k V$, $\mathfrak{B} \in \bigwedge^l V$ the generalized inner product is defined by

$$\mathfrak{A} \cdot \mathfrak{B} := [\mathfrak{A}\mathfrak{B}]_{|k-l|},$$

where $[\cdot]_m$, $m \in \mathbb{N}$ denotes the grade-m part.

The outer Product The outer or exterior product for two vectors is given by

$$\mathfrak{a} \wedge \mathfrak{b} := \frac{1}{2}(\mathfrak{a}\mathfrak{b} - \mathfrak{b}\mathfrak{a}). \tag{1.28}$$

This product can also be generalized to blades, cf. [36]. For $\mathfrak{A} \in \bigwedge^k V$, $\mathfrak{B} \in \bigwedge^l V$ the exterior product is defined by

$$\mathfrak{A} \wedge \mathfrak{B} := [\mathfrak{A}\mathfrak{B}]_{|k+l|}.$$

The geometric Product With the inner (1.27) and the exterior product (1.28) the geometric product of vectors can be written in the following form

$$\mathfrak{a}\mathfrak{b} = \mathfrak{a} \cdot \mathfrak{b} + \mathfrak{a} \wedge \mathfrak{b}. \tag{1.29}$$

Definition 1.12. *A k-blade is the k-fold exterior product of vectors or grade-1 elements $\mathfrak{v} \in \bigwedge^1 V$. Therefore, a k-blade can be written as*

$$\mathfrak{A} = \mathfrak{a}_1 \wedge \mathfrak{a}_2 \wedge \ldots \wedge \mathfrak{a}_k =: \bigwedge_{i=1}^{k} \mathfrak{a}_i.$$

Grade-k elements that are k-blades are called simple *or* decomposable. *A k-blade that squares to zero is called a* null *k-blade.*

Remark 1.5. *Not every grade-k element is also a k-blade. The two-blade $e_{12} + e_{13}$ can be written as $e_1 \wedge (e_2 + e_3)$. As example for a grade-2 element that is not a two-blade we take $e_{12} + e_{34}$.*

For treating geometric entities within geometric algebra context the definition of the *inner product null space* and its dual the *outer product null space* is needed.

Definition 1.13. *The* inner product null space *(IPNS) of a k-blade* $\mathfrak{A} \in \bigwedge^k V$, *is defined by*

$$\mathbb{NI}(\mathfrak{A}) := \left\{ \mathfrak{v} \in \bigwedge^1 V : \mathfrak{v} \cdot \mathfrak{A} = 0 \right\}.$$

Moreover, the outer product null space *(OPNS) of a k-blade* $\mathfrak{A} \in \bigwedge^k V$ *is defined by*

$$\mathbb{NO}(\mathfrak{A}) := \left\{ \mathfrak{v} \in \bigwedge^1 V : \mathfrak{v} \wedge \mathfrak{A} = 0 \right\}.$$

The IPNS and the OPNS of k-blades can be used to describe subspaces of the vector space V, see [53]. Furthermore, we have the property of the outer and inner product null space

$$\mathbb{NI}(\mathfrak{a} \wedge \mathfrak{b}) = \mathbb{NI}(\mathfrak{a}) \cap \mathbb{NI}(\mathfrak{b}), \quad \mathbb{NO}(\mathfrak{a} \wedge \mathfrak{b}) = \mathbb{NO}(\mathfrak{a}) \oplus \mathbb{NO}(\mathfrak{b}). \tag{1.30}$$

Table 1.1: Clifford algebras $\mathcal{Cl}_{(p,q,0)}$, see [55].

	q→ 0	1	2	3	4	5	6	7
p↓ 0	$\mathbb{R}$	$\mathbb{C}$	$\mathbb{H}$	$^2\mathbb{H}$	$\mathbb{H}(2)$	$\mathbb{C}(4)$	$\mathbb{R}(8)$	$^2\mathbb{R}(8)$
1	$^2\mathbb{R}$	$\mathbb{R}(2)$	$\mathbb{C}(2)$	$\mathbb{H}(2)$	$^2\mathbb{H}(2)$	$\mathbb{H}(4)$	$\mathbb{C}(8)$	$\mathbb{R}(16)$
2	$\mathbb{R}(2)$	$^2\mathbb{R}(2)$	$\mathbb{R}(4)$	$\mathbb{C}(4)$	$\mathbb{H}(4)$	$^2\mathbb{H}(4)$	$\mathbb{H}(8)$	$\mathbb{C}(16)$
3	$\mathbb{C}(2)$	$\mathbb{R}(4)$	$^2\mathbb{R}(4)$	$\mathbb{R}(8)$	$\mathbb{C}(8)$	$\mathbb{H}(8)$	$^2\mathbb{H}(8)$	$\mathbb{H}(16)$
4	$\mathbb{H}(2)$	$\mathbb{C}(4)$	$\mathbb{R}(8)$	$^2\mathbb{R}(8)$	$\mathbb{R}(16)$	$\mathbb{C}(16)$	$\mathbb{H}(16)$	$^2\mathbb{H}(16)$
5	$^2\mathbb{H}(2)$	$\mathbb{H}(4)$	$\mathbb{C}(8)$	$\mathbb{R}(16)$	$^2\mathbb{R}(16)$	$\mathbb{R}(32)$	$\mathbb{C}(32)$	$\mathbb{H}(32)$
6	$\mathbb{H}(4)$	$^2\mathbb{H}(4)$	$\mathbb{H}(8)$	$\mathbb{C}(16)$	$\mathbb{R}(32)$	$^2\mathbb{R}(32)$	$\mathbb{R}(64)$	$\mathbb{C}(64)$
7	$\mathbb{C}(8)$	$\mathbb{H}(8)$	$^2\mathbb{H}(8)$	$\mathbb{H}(16)$	$\mathbb{C}(32)$	$\mathbb{R}(64)$	$^2\mathbb{R}(64)$	$\mathbb{R}(128)$

Table 1.1 shows the matrix representation of the non-degenerate Clifford algebras $\mathcal{Cl}_{(p,q,0)}$. In this table $\mathbb{R}(n)$ denotes the $n \times n$-matrices over $\mathbb{R}$ and $^2\mathbb{R}(n)$ denotes $\mathbb{R}(n) \times \mathbb{R}(n)$.

1.3 Geometric Algebras, Clifford Algebras

1.3.3 Pin and Spin Groups

Not every element of a Clifford algebra has an inverse element. In general there are zero divisors, this means for two non zero elements $\mathfrak{A}, \mathfrak{B} \in \mathcal{Cl}_{(p,q,r)}$ the geometric product $\mathfrak{AB}$ is zero.

Example 1.4. *A simple example for a zero divisor is given by an element* $\mathfrak{a} = ae_1 + ae_2 \in \mathcal{Cl}_{(1,1,0)}$, $a \in \mathbb{R}$. *This Clifford algebra has the standard basis* $\{e_0, e_1, e_2, e_{12}\}$ *with* $e_1^2 = 1$, $e_2^2 = -1$ *and* $e_{12}^2 = 1$. *We calculate*

$$\mathfrak{a}^2 = (ae_1 + ae_2)^2 = a^2 e_1^2 + a^2 e_{12} + a^2 e_{21} + a^2 e_2^2 = 0.$$

Note that $\mathcal{Cl}_{(1,1,0)} \cong \mathbb{R}^{2 \times 2}$. We give an isomorphism by its action on the standard basis, cf. [3].

$$e_0 \mapsto \begin{pmatrix} 1 & 0 \\ 0 & 1 \end{pmatrix}, \quad e_1 \mapsto \begin{pmatrix} 1 & 0 \\ 0 & -1 \end{pmatrix}, \quad e_2 \mapsto \begin{pmatrix} 0 & -1 \\ 1 & 0 \end{pmatrix}, \quad e_{12} \mapsto \begin{pmatrix} 0 & 1 \\ 1 & 0 \end{pmatrix}.$$

The inverse element of a versor $\mathfrak{V}$ is defined by $\mathfrak{V}^{-1} = \frac{\mathfrak{V}^*}{N(\mathfrak{V})}$ with $N(\mathfrak{V}) = \mathfrak{VV}^*$. The map $N : \mathcal{Cl}_{(p,q,r)} \to \mathcal{Cl}_{(p,q,r)}$ is called the *norm* of the Clifford algebra. For general multivectors $\mathfrak{M} \in \mathcal{Cl}_{(p,q,r)}$ inverse elements exist and are defined by the relation $\mathfrak{M}\mathfrak{M}^{-1} = \mathfrak{M}^{-1}\mathfrak{M} = 1$, but the computation is more difficult. This is discussed in section 1.3.4.

Clifford Group Invertible elements are called *units*. We denote the set of units of a Clifford algebra by $\mathcal{Cl}^\times_{(p,q,r)}$. For a general ring $\mathcal{R}$, we denote the set of units by $\mathcal{R}^\times$. With respect to the geometric product the units form a group. For a given Clifford algebra the *Clifford group* is the group

$$\Gamma(\mathcal{Cl}_{(p,q,r)}) := \left\{ \mathfrak{g} \in \mathcal{Cl}^\times_{(p,q,r)} \,|\, \alpha(\mathfrak{g})\mathfrak{v}\mathfrak{g}^* \in V \, \forall \mathfrak{v} \in V \right\}. \tag{1.31}$$

It is not obvious that $\Gamma(\mathcal{Cl}_{(p,q,r)})$ is a group. For a proof we refer to [21]. Elements of the Clifford group are denoted by the small gothic letter $\mathfrak{g}$ if not stated otherwise. We are interested in special subgroups of the group of units.

The Pin Group   The Pin group is defined by

$$\mathrm{Pin}_{(p,q,r)} := \{\mathfrak{g} \in \mathcal{C}\ell_{(p,q,r)} \,|\, N(\mathfrak{g}) = \pm 1,\ \alpha(\mathfrak{g})\mathfrak{v}\mathfrak{g}^* \in V\ \forall \mathfrak{v} \in V\,\}. \qquad (1.32)$$

The action of the so called *sandwich operator* $\alpha(\mathfrak{g})\mathfrak{v}\mathfrak{g}^*$ with $\mathfrak{g} \in \mathrm{Pin}_{(p,q,r)}$ applied to vectors does not change the scalar product of two vectors. This can be verified easily by direct calculation. Note that the condition $N(\mathfrak{g}) = 1$ guarantees that $\mathfrak{g}$ is a unit. Let $\mathfrak{a}, \mathfrak{b} \in \bigwedge^1 V$ and $\mathfrak{g} \in \mathrm{Pin}(p,q,r)$. Furthermore, let $\mathfrak{a}' = \alpha(\mathfrak{g})\mathfrak{a}\mathfrak{g}^*$ and $\mathfrak{b}' = \alpha(\mathfrak{g})\mathfrak{b}\mathfrak{g}^*$. The scalar product can be expressed in terms of the geometric product as in Eq. (1.27). We use this equation for the transformed vectors $\mathfrak{a}'$ and $\mathfrak{b}'$ and find

$$\begin{aligned}\mathfrak{a}' \cdot \mathfrak{b}' &= \frac{1}{2}\left(\alpha(\mathfrak{g})\mathfrak{a}\mathfrak{g}^*\alpha(\mathfrak{g})\mathfrak{b}\mathfrak{g}^* + \alpha(\mathfrak{g})\mathfrak{b}\mathfrak{g}^*\alpha(\mathfrak{g})\mathfrak{a}\mathfrak{g}^*\right) \\ &= \frac{1}{2}\alpha(\mathfrak{g})\left[\mathfrak{a}\mathfrak{g}^*\alpha(\mathfrak{g})\mathfrak{b} + \mathfrak{b}\mathfrak{g}^*\alpha(\mathfrak{g})\mathfrak{a}\right]\mathfrak{g}^*.\end{aligned}$$

The main involution applied to a Pin group element $\alpha(\mathfrak{g})$ is either $\mathfrak{g}$ or $-\mathfrak{g}$. This follows, because a Pin group element $\mathfrak{g}$ can be written as $\mathfrak{g} = \mathfrak{g}_1 \ldots \mathfrak{g}_n$, where $\mathfrak{g}_1, \ldots, \mathfrak{g}_k$, $k \leq n$ are grade-1 elements. The main involution applied to a grade-1 element $\mathfrak{g}_1 \in \bigwedge^1 V$ equals $-\mathfrak{g}_1$. Therefore, we conclude that $\mathfrak{g}^*\alpha(\mathfrak{g}) = \alpha(\mathfrak{g})\mathfrak{g}^*$ is either 1 or -1 depending on if k is even or odd. Therefore, the term in square brackets is a scalar and consequently

$$\mathfrak{a}' \cdot \mathfrak{b}' = \frac{1}{2}\alpha(\mathfrak{g})\mathfrak{g}^*\left[\mathfrak{g}^*\alpha(\mathfrak{g})(\mathfrak{a}\mathfrak{b} + \mathfrak{b}\mathfrak{a})\right].$$

Thus, we get

$$\mathfrak{a}' \cdot \mathfrak{b}' = \frac{1}{2}(\alpha(\mathfrak{g})\mathfrak{g}^*)(\mathfrak{g}^*\alpha(\mathfrak{g}))\,(\mathfrak{a}\mathfrak{b} + \mathfrak{b}\mathfrak{a}) = \frac{1}{2}(\mathfrak{a}\mathfrak{b} + \mathfrak{b}\mathfrak{a}) = \mathfrak{a} \cdot \mathfrak{b}.$$

Linear transformations of $\bigwedge^1 V$ that preserve distances and angles, *i.e.*, the scalar product, are elements of the orthogonal group $O(p,q,r)$. In fact, $\mathrm{Pin}(p,q,r)$ is a double cover of the orthogonal group $O(p,q,r)$, see [21].

1.3 Geometric Algebras, Clifford Algebras

Remark 1.6. *Usually, the sandwich operator for elements of the Clifford group $\Gamma_{(p,q,r)}$ is defined by $\alpha(\mathfrak{g})\mathfrak{v}\mathfrak{g}^{-1}$. For elements of the Pin group the operator simplifies because for $\mathfrak{g} \in \operatorname{Pin}_{(p,q,r)}$ we have*

$$\mathfrak{g}^{-1} = \frac{\mathfrak{g}^*}{\mathfrak{g}\mathfrak{g}^*} = \mathfrak{g}^*.$$

In the homogeneous Clifford Algebra models we use $\mathfrak{g}^*$ because multiplication with a homogeneous factors does not change the geometric meaning of the sandwich operator.

Remark 1.7. *The mapping $\mathfrak{v} \mapsto \alpha(\mathfrak{g})\mathfrak{v}\mathfrak{g}^*$ for $\mathfrak{g} \in \operatorname{Pin}(p,q,r) \cap \bigwedge^1 V$ is a reflection in the hyperplane perpendicular to $\mathbb{NO}(\mathfrak{v})$ with respect to ρ through the origin. The composition of two reflections is again an element of the Pin group.*

The Spin Group The second important subgroup of the group of units is the *Spin group* defined via

$$\operatorname{Spin}_{(p,q,r)} := \left\{ \mathfrak{g} \in \mathcal{C}\ell^+_{(p,q,r)} \,\middle|\, N(\mathfrak{g}) = \pm 1,\ \mathfrak{g}\mathfrak{v}\mathfrak{g}^* \in V\ \forall \mathfrak{v} \in V \right\} \qquad (1.33)$$
$$= \operatorname{Pin}_{(p,q,r)} \cap \mathcal{C}\ell^+_{(p,q,r)}.$$

Note that the main involution α has no effect on the even subalgebra, and therefore, can be neglected.

Remark 1.8. *The Spin group is a subgroup of the Pin group. Thus, the scalar product of vectors is preserved under the action of the Spin group. Spin group elements are generated by pairs of reflections, so they are rotations.*

The Spin group for degenerate Clifford algebras as in the case of Euclidean geometry are semi-direct products of Spin groups for the non-degenerate part and an additive matrix group, see [15]. Later we will construct the Clifford algebra for the three-dimensional Euclidean space.

Remark 1.9. *Due to their definitions the Pin and the Spin group consist of two connected components. The connected components are* given by $N(\mathfrak{g}) = 1$ and $N(\mathfrak{g}) = -1$. We distinguish the two connected components by

$$Pin^+_{(p,q,r)} := \{\mathfrak{g} \in \mathcal{Cl}_{(p,q,r)} \,|\, N(\mathfrak{g}) = +1,\ \alpha(\mathfrak{g})\mathfrak{v}\mathfrak{g}^* \in V\ \forall \mathfrak{v} \in V\},$$
$$Pin^-_{(p,q,r)} := \{\mathfrak{g} \in \mathcal{Cl}_{(p,q,r)} \,|\, N(\mathfrak{g}) = -1,\ \alpha(\mathfrak{g})\mathfrak{v}\mathfrak{g}^* \in V\ \forall \mathfrak{v} \in V\}.$$

For the connected components of the Spin group we use the same notation.

1.3.4 Matrix Representation of Clifford Algebras

In this section we follow [50]. Any multivector of a Clifford algebra over an n-dimensional vector space $n \in \mathbb{N}$ can be represented by a vector $v \in \mathbb{R}^{2^n}$. Then the geometric product of two algebra elements $\mathfrak{A}\mathfrak{B} = \mathfrak{C}$ can be written as product of a matrix and a vector

$$C = [A^+]\,B, \quad C = [B^-]\,A, \quad [A^+],[B^-], \tag{1.34}$$

where $[A^+]$ and $[B^-]$ are $2^n \times 2^n$ matrices with real entries and $A, B, C \in \mathbb{R}^{2^n}$. The columns of the matrix $[A^+]$ are defined by the products

$$\mathfrak{a}e_0, \mathfrak{a}e_1, \ldots, \mathfrak{a}e_n, \mathfrak{a}e_{12}, \ldots \mathfrak{a}e_{(n-1)n}, \ldots, \mathfrak{a}e_{1\ldots n}.$$

Note that these columns are ordered from right to left to match the chosen ordering of the vector representation for the Clifford algebra. To obtain the columns of the matrix $[B^-]$ we multiply b from the left side with each basis element

$$e_0\mathfrak{b}, e_1\mathfrak{b}, \ldots, e_n\mathfrak{b}, e_{12}\mathfrak{b}, \ldots e_{(n-1)n}\mathfrak{b}, \ldots, e_{1\ldots n}\mathfrak{b}$$

again with the same ordering.

Remark 1.10. *With the matrix representation it is possible to compute the inverse element for an arbitrary multivector. In order to achieve this, we express the geometric product as product of a matrix with a vector. If the $2^n \times 2^n$ matrix is invertible, the inverse algebra element corresponds to the inverse matrix. For increasing vector space dimension this calculation can be extremely expensive.*

Remark 1.11. *The matrix representation of the dual quaternion product that was determined in section 1.1.2 is computed in the same way.*

1.3.5 Linear Transformation of the Vector Space

The action of the sandwich operator applied to vectors can be written as a linear transformation of the vector space. Therefore, we rewrite the sandwich product of an arbitrary versor $\mathfrak{g} \in \mathcal{Cl}_{(p,q,r)}$ applied to a vector $\mathfrak{v} \in \bigwedge^1 V$ as product of a matrix with a vector

$$\alpha(\mathfrak{g})\mathfrak{v}\mathfrak{g}^* = \mathfrak{v}' \Rightarrow \mathrm{K} \cdot v = v',$$

where $v, v' \in \mathbb{R}^n$ and $\mathrm{K} \in \mathbb{R}^{n \times n}$ is a linear mapping. This procedure can be applied to every subspace of the Clifford Algebra $\mathcal{Cl}_{(p,q,r)}$. We illustrate this at hand of the following example.

Example 1.5. *The Clifford algebra $\mathcal{Cl}_{(0,2,0)} \cong \mathbb{H}$, see Ex. 1.1 serves as an example. Hence, we take two general elements of the form*

$$\mathfrak{g} = a_0 e_0 + a_1 e_2 + a_2 e_2 + a_3 e_{12},$$
$$\mathfrak{v} = x_0 e_0 + x_1 e_2 + x_2 e_2 + x_3 e_{12},$$

where $\mathfrak{g}$ is a versor and $\mathfrak{v}$ is arbitrary. We apply the sandwich operator to $\mathfrak{v}$:

$$\begin{aligned}\alpha(\mathfrak{g})\mathfrak{v}\mathfrak{g}^* =\ & ((a_0^2 - a_2^2 + a_3^2 - a_1^2)x_0 + 2(a_0 a_1 + a_3 a_2)x_1 + 2(a_0 a_2 - a_1 a_3)x_2)e_0 \\ & + (2(a_3 a_2 - a_0 a_1)x_0 + (a_0^2 - a_1^2 + a_2^2 - a_3^2)x_1 - 2(a_0 a_3 + a_1 a_2)x_2)e_1 \\ & + (-2(a_1 a_3 x_0 + a_0 a_2)x_0 + 2(a_0 a_3 - a_1 a_2)x_1 \\ & + (a_1^2 - a_2^2 - a_3^2 + a_0^2)x_2)e_2 + (a_3^2 + a_0^2 + a_1^2 + a_2^2)x_3 e_{12}.\end{aligned}$$

If we now write this operation as a product of a matrix with a vector, we obtain $v' = \mathrm{K} \cdot v$, with $v = (x_0, x_1, x_2, x_3)^\mathrm{T} \in \mathbb{R}^4$ and

$$\mathrm{K} = \begin{pmatrix} a_3^2 - a_1^2 + a_0^2 - a_2^2 & 2a_0 a_1 + 2a_3 a_2 & 2a_0 a_2 - 2a_1 a_3 & 0 \\ -2a_0 a_1 + 2a_3 a_2 & a_2^2 - a_1^2 - a_3^2 + a_0^2 & -2a_0 a_3 - 2a_1 a_2 & 0 \\ -2a_1 a_3 - 2a_0 a_2 & -2a_1 a_2 + 2a_0 a_3 & a_1^2 - a_2^2 - a_3^2 + a_0^2 & 0 \\ 0 & 0 & 0 & a_0^2 + a_1^2 + a_2^2 + a_3^2 \end{pmatrix}.$$

If $\mathfrak{g} \in \mathrm{Pin}_{(0,2,0)}$ it yields $a_0^2 + a_1^2 + a_2^2 + a_3^2 = 1$. The matrix K represents a rotation, where x_3 contains the homogeneous component, see Eq. (1.4). This means that in the projective space $\mathbb{P}^3(\mathbb{R})$ the hyperplane $x_3 = 0$ is identified as ideal plane and that the affine part is obtained as $\mathbb{P}^3(\mathbb{R}^3 \times 1)$ in $\mathbb{P}^3(\mathbb{R}^3 \times \mathbb{R})$, see [31].

1.4 The Homogeneous Model

In this section we follow [19, 27] and [48] and construct the Clifford algebra over a projective space $\mathbb{P}^n(\mathbb{R})$ modelled by an $(n + 1)$-dimensional vector space V endowed with a quadratic form. The resulting model is called the *homogeneous model*. In this model, it is possible to describe affine subspaces, *i.e.*, translated subspaces by algebra elements respectively their IPNS or OPNS. Hence, we disengage ourselves from the origin. Points of $\mathbb{P}^n(\mathbb{R})$ are represented by vectors respectively one-dimensional subspaces of V. This enables us to interpret the grade-1 subspace $\bigwedge^1 V$ as a copy of a projective space. The homogeneous model is nothing else but the geometric algebra representation of homogeneous coordinates from projective geometry. Meet and join operations are now available for affine subspaces.

1.4.1 The Construction

As suggested in the preliminaries of this section we use a vector space of dimension $n+1$ to define the geometric algebra corresponding to the homogeneous model for an n-dimensional vector space. This means we add one generator $e_\sim$ to the standard basis of a given vector space V. Later in chapter 3, we see that the choice of the square ($e_\sim^2 = 1, -1$, or 0) will determine a Cayley-Klein geometry. Furthermore, we define this direction to be orthogonal to all other directions of V: $e_i \cdot e_\sim = 0$ for all $i = 1, \ldots, n$ with respect to the quadratic form ρ. Points of the space are then represented by grade-1 elements, *i.e.*, vectors, directed lines by two-blades and so on. The occurring geometric entities represented by subspaces are described in the Grassmann or exterior algebra. Let $P = (p_1, \ldots, p_n)^\mathrm{T} \in V$ be a point. It has homogeneous coordinates $\underline{P} = (1, p_1, \ldots, p_n)^\mathrm{T} \mathbb{R}$ and as an element of the algebra, it reads $\mathfrak{p} = e_\sim + p_1 e_1 + \ldots + p_n e_n$. With the wedge product it is possible to define algebra elements corresponding to the join of other algebra elements. A line is the wedge product of two points that are represented by vectors in the homogeneous Clifford algebra model, and a plane can be described as the wedge product of three non-collinear points. These subspaces can be described with Grassmann coordinates.

1.4.2 Exterior Algebra

The *exterior algebra* or *Grassmann algebra* $\bigwedge V$ over a real vector space V assigns points to subspaces of projective spaces. Therefore, we take a closer look on the wedge product that we have already met in section 1.3. In the following we assume that one-dimensional subspaces in V represent points of the projective space $\mathbb{P}(V)$. We denote the vector space with V and its standard basis with $\{e_1, \ldots, e_n\}$.

Definition 1.14. *For an n-dimensional real vector space V the Grassmann algebra $\bigwedge V$ is a Clifford algebra with fully degenerate quadratic form. This means the matrix Q of the quadratic form is the $n \times n$ zero matrix. Multiplication in $\bigwedge V$ is denoted by the wedge product $\wedge$ and sometimes referred to as* exterior product.

From the properties of the geometric product we have for $\mathfrak{X}, \mathfrak{Y} \in \bigwedge V$

$$\mathfrak{X} \wedge \mathfrak{Y} + \mathfrak{Y} \wedge \mathfrak{X} = 0.$$

The exterior product is an alternating multilinear operation. In fact, the exterior product is the geometric product for this special algebra. For vectors $\mathfrak{v}_1, \mathfrak{v}_2, \ldots \mathfrak{v}_k$ it is zero if, and only if, they are linearly dependent. If $\{e_1, \ldots, e_n\}$ is a basis of V, then $e_{i_1 i_2 \ldots i_k}$, $0 \leq k \leq n$, $i_1 < i_2 < \ldots < i_n$ is a basis for $\bigwedge V$, for $k = 0$ we define $e_{i_0} := e_0$, i.e., the scalar part. Furthermore, the Grassmann algebra possesses the same graduation as every universal Clifford algebra. We have the following direct sum

$$\bigwedge V = \bigwedge^0 V \oplus \bigwedge^1 V \oplus \ldots \oplus \bigwedge^n V.$$

The dimension of $\bigwedge^k V$ equals $\binom{n}{k}$, and therefore, the dimension of $\bigwedge V$ equals 2^n. The subspace $\bigwedge^0 V$ is identified with the real numbers, i.e., scalars. Elements of $\bigwedge^1 V$ are identified with elements of V respectively $\mathbb{P}^n(\mathbb{R})$. Elements from $\bigwedge^k V$ are called *k-vectors*. We recall a theorem from [56] without proof.

Theorem 1.1. *For $\mathcal{S} \in \bigwedge^k V$ the set of all $\mathfrak{a} \in \bigwedge^1 V$ such that $\mathcal{S} \wedge \mathfrak{a} = 0$ is a linear subspace denoted by $L(\mathcal{S})$. If $\mathcal{S}$ is non-zero and simple with $\mathcal{S} = \mathfrak{v}_1 \wedge \ldots \wedge \mathfrak{v}_k$, then $L(\mathcal{S})$ equals the span $[\mathfrak{v}_1, \ldots, \mathbf{v}_k]$. For simple $\mathcal{S}_1, \mathcal{S}_2$ the spaces $L(\mathcal{S}_1)$ and $L(\mathcal{S}_2)$ are equal if, and only if, $\mathcal{S}_1$ is a scalar multiple of $\mathcal{S}_2$.*

Th. 1.1 gives a one-to-one correspondence between linear subspaces $L(\mathcal{S})$ and one-dimensional subspaces in $\bigwedge^k V$ spanned by simple elements. We define:

Definition 1.15. *For a k-dimensional linear subspace $\mathcal{S}$ spanned by $[\mathfrak{v}_1, \ldots, \mathfrak{v}_k]$ we compute*

$$\mathfrak{v}_1 \wedge \mathfrak{v}_2 \wedge \ldots \wedge \mathfrak{v}_k = \sum_{i_1 < \ldots < i_k} v_{i_1 \ldots i_k} e_{i_1} \wedge \ldots \wedge e_{e_k}.$$

The coefficients $v_{i_1 \ldots i_k}$ are called Grassmann coordinates *of $\mathcal{S}$.*

Note that by Th. 1.1 Grassmann coordinates are defined only up to a constant scalar factor. Therefore, Grassmann coordinates can be interpreted as homogeneous coordinates. Furthermore, not every element of $\bigwedge^k V$ is simple, *i.e.*, decomposable. Therefore, we need more relations to describe simple elements. Here we do not investigate these relations, the interested reader is referred to [56]. To clarify these concepts, we give an example.

Example 1.6. *Let $V = \mathbb{R}^4$ be the vector space model for $\mathbb{P}^3(\mathbb{R})$. Moreover, two points $P_1 = (1,3,2,0)^T \mathbb{R}$ and $P_2 = (1,0,2,1)^T \mathbb{R}$ are given in Grassmann algebra representation by $\mathfrak{p}_1 = e_0 + 3e_1 + e_2$ and $\mathfrak{p}_2 = e_0 + 2e_2 + e_3$. Now we calculate the exterior product*

$$(e_0 + 3e_1 + e_2) \wedge (e_0 + 2e_2 + e_3) = -3(e_0 \wedge e_1) + (e_0 \wedge e_2) + (e_0 \wedge e_3)$$
$$+ 6(e_1 \wedge e_2) + 3(e_1 \wedge e_3) + (e_2 \wedge e_3).$$

These Grassmann coordinates can now be listed as $(-3, 1, 1, 6, 3, 1)$. The coordinates are Plücker coordinates if we order them in the right way and use the historic convention that the coefficient of $e_3 \wedge e_1$ is used. The line spanned by P_1 and P_2 has the Plücker coordinates $(-3, 1, 1, 1-3, 6)$. Here, the Plücker condition can be checked easily. To test if an arbitrary point P_3 with algebra representation $\mathfrak{p}_3$ lies on the line $[P_1, P_2]$ we just have to apply the exterior product one more time. If the result is zero the point is incident with the line.

The example above also shows that in the case of Plücker coordinates simple elements satisfy an additional condition. In this case it is the condition that they all have to lie on a hyperquadric $M_2^4 \subset \mathbb{P}^5(\mathbb{R})$, see section 1.2.1.

Dual Exterior Algebra In some cases it is necessary to construct a homogeneous Clifford algebra model with the use of the dual exterior algebra. This is the reason why we give a short sketch for the construction of the dual of an exterior algebra. All that has to be done is to do the same construction on the dual vector space V^*, i.e., the vector space of all linear-forms on V. The dual exterior algebra is then the algebra of the k-multilinear forms and it is naturally isomorphic to $\bigwedge V$ again. In $\bigwedge V^*$ we can also find the subspace structure of $\mathbb{P}^n(\mathbb{R})$ again. The difference is that elements from $\bigwedge^1 V$ represent hyperplanes, elements from $\bigwedge^2 V$ hyper lines. Furthermore, elements from $\bigwedge^{n-1} V$ represent points and so on. All in all the structure of $\bigwedge V^*$ is isomorphic to the structure of $\bigwedge V$ turned on its head. Later we will define this isomorphism and use it to express the meet operation with the wedge product.

1.4.3 Homogeneous Model via projective Grassmann Algebra

From section 1.3 we know how to define a Clifford algebra. A Clifford algebra and a Grassmann algebra have the same graduation. If we now equip $\bigwedge^1 V$ with an inner product, we define a geometric product on vectors for $\bigwedge V$. In a dual way we could define a geometric product on the $\bigwedge^{n-1} V$ space for $\bigwedge V^*$. In [28] this construction is presented in detail. The isomorphism presented now can also be found there. In the literature duality in a Clifford algebra is often defined through multiplication with the pseudoscalar. When we deal with degenerate Clifford algebras this definition is not adequate. Therefore, we define duality by the so called *Poincaré isomorphism*. This isomorphism can be used to define projective meet operations.

Definition 1.16. *The isomorphism*

$$J : \mathcal{C}\ell_{(p,q,r)} \to \mathcal{C}\ell_{(p,q,r)},$$
$$e_{i_1 \ldots i_k} \mapsto e_{I \setminus \{i_1, \ldots, i_k\}}, 0 \leq i_1 < \ldots < i_k \leq n,$$

where I is the ordered set $\{0, \ldots, n\}$ and $n = p + q + r$ the dimension of the vector space is called the Poincaré duality. Grade-k elements are mapped to grade-$(n-k)$ elements.

Example 1.7. *If we apply this mapping to a grade-2 element* $\mathfrak{p} = x_0 e_{12} + x_1 e_{23} + x_2 e_{13}$ *of* $\mathcal{C}\ell_{(2,0,1)}$ *we get a grade-1 element* $J(\mathfrak{p}) = x_0 e_3 + x_1 e_1 + x_2 e_2$. *The scalar product* (1.27) *results in*

$$J(\mathfrak{p}) \cdot J(\mathfrak{p}) = x_1^2 + x_2^2.$$

This expression belongs to the norm of vectors in the Euclidean plane. Note, that in this case

$$J(\mathfrak{p}) J(\mathfrak{p}) = J(\mathfrak{p}) \cdot J(\mathfrak{p}),$$

since the exterior product of an element with itself vanishes. It can be interpreted as the squared distance between the origin and the hyperplane defined by $\mathbb{NO}(J(\mathfrak{p}))$. *Therefore, the geometric entity corresponding to the outer product null space of* $J(\mathfrak{p})$ *can be interpreted as line, i.e., a hyperplane in* $\mathbb{P}^2(\mathbb{R})$.

This example shows that for Euclidean geometry we have to define the inner product for hyperplanes and therefore, we have to use the dual Grassmann algebra. A full description of the homogeneous model lies beyond the scope of this work. The interested reader is invited to consult the cited references. In chapter 3 we take a closer look on special homogeneous Clifford algebra models to construct kinematic mappings.

1.5 The Conformal Model

The next step was taken by D. HESTENES, see [34]. He designed a geometric algebra that is capable to describe conformal transformations with its Pin group. Furthermore, a lot more geometric objects find their analogue as algebra elements. There is a huge amount of literature on conformal geometric algebra, abbreviated by CGA, and the interested reader is referred to [18, 19, 35, 34] or [46].

To arrive at the homogeneous model we had to add one extra dimension to get rid of of the origin. In the homogeneous Clifford algebra model we are able to describe translations by Spin group elements. The idea to model conformal geometry is to add an extra dimension with special properties to the homogeneous model.

1.5.1 Construction of Conformal Geometric Algebra

We denote the n-dimensional Euclidean space by $\mathbb{E}^n$. The associated quadratic form shall be the standard scalar product. The basis vector space is the Cartesian one with two additional generators e_+ and e_- squaring to 1 and -1 as prescribed by the index of the corresponding additional generator. This means that we get a *Minkowski algebra* and the quadratic form ρ has signature $(n+1,1,0)$. For our considerations it is more convenient to use a different basis for the quadratic space $\mathbb{R}^{(n+1,1)}$, i.e., $\mathbb{R}^{n+2}$ is equipped with the quadratic form ρ. We introduce the two isotropic vectors

$$o = \frac{1}{2}(e_- + e_+), \quad \infty = (e_- - e_+).$$

Note that these two generators square to zero. In geometric algebra context isotropic vectors are called *null vectors*. Sometimes in the literature o is denoted by no, for " *null vector representing the origin* " and ∞ by ni, for " *null vector representing infinity* ". Especially for $n = 3$ the multiplication table is given by

	o	e_1	e_2	e_3	∞
o	0	0	0	0	-1
e_1	0	1	0	0	0
e_2	0	0	1	0	0
e_3	0	0	0	1	0
∞	-1	0	0	0	0

There are also other possibilities to define the basis of $\mathbb{R}^{(4,1)}$. This algebra can be thought of as the homogeneous model with one additional restriction in the ∞-direction. The linear subspaces of $\mathbb{R}^{n+1}$ that model affine subspaces in $\mathbb{R}^n$ are intersected with an $(n+1)$-dimensional paraboloid, called the *horosphere*, see [20, 35]. The intersection is then projected on the n-dimensional hyperplane spanned by $[e_1, \ldots, e_n]$. The scalar product in conformal geometric algebra (abbreviated by CGA) helps to give an analogue to the Euclidean distance. Therefore, points are modelled by null vectors, because they have to satisfy $\mathfrak{p} \cdot \mathfrak{p} = \mathfrak{p}\mathfrak{p} = 0$, since $\mathfrak{p} \wedge \mathfrak{p} = 0$. For a point $P \in \mathbb{R}^n$ the corresponding algebra element has the form

$$\mathfrak{p} = o + p + \frac{1}{2}p^2 \infty,$$

where $p = p_1 e_1 + \ldots + p_n e_n$ is a vector in $\mathbb{R}^n$. With this definition it easy to check that the scalar product of two such elements $\mathfrak{p}$ and $\mathfrak{q}$ evaluates to

$$\mathfrak{p} \cdot \mathfrak{q} = (o+p+\frac{1}{2}p^2\infty) \cdot (o+q+\frac{1}{2}q^2\infty) \qquad (1.35)$$
$$= -\frac{1}{2}q^2 + p \cdot q - \frac{1}{2}p^2 = -\frac{1}{2}(q-p)^2.$$

With Eq. (1.35) we see that points are described by null vectors, since $\mathfrak{p} \cdot \mathfrak{p} = \mathfrak{p}^2 = 0$ and $\mathfrak{p} \wedge \mathfrak{p} = 0$. Furthermore, it now becomes clear that the Euclidean distance is integrated in the algebra by the scalar product. Note that the o-component can be understood as the homogeneous factor. This factor can be extracted from a point by

$$-\infty \cdot \mathfrak{p} = -\infty \cdot \alpha(o+p+\frac{1}{2}p^2\infty) = \alpha.$$

Furthermore, a point is called *normalized* if the homogeneous factor is equal to 1. We have to remember that if we want to calculate Euclidean distances with the scalar product. The inner product null space and the outer product null space with respect to the embedding is called the geometric inner and geometric outer product null space. It depends on the embedding of the base space into a higher dimensional space. We denote the embedding by η. Especially, non-linear embeddings give rise to interesting geometric entities. Therefore, we recall the definition, see [53].

Definition 1.17. *The* geometric inner product null space *(GIPNS) and dual the* geometric outer product null space *(GOPNS) of a k-blade $\mathfrak{A} \in \bigwedge^k V$ is defined as*

$$\mathbb{NI}_G(\mathfrak{A}) := \left\{ x \in \mathbb{R}^{(r,s)} : \eta(x) \cdot \mathfrak{A} = 0 \right\},$$
$$\mathbb{NO}_G(\mathfrak{A}) := \left\{ x \in \mathbb{R}^{(r,s)} : \eta(x) \wedge \mathfrak{A} = 0 \right\}.$$

The mapping $\eta : \bigwedge^1 \mathbb{R}^{(r,s)} \to \mathbb{X}$, where $\mathbb{X} \subseteq \bigwedge^1 \mathbb{R}^{(p,q)}$, with $r+s \leq p+q$ embeds the base space into $\bigwedge^1 \mathbb{R}^{(p,q)}$.

Note that we mean the Poincaré duality when we talk about duality in geometric algebra context.

1.5 The Conformal Model

Remark 1.12. *In three-dimensional conformal geometric algebra the embedding η of three-dimen-sional space $\mathbb{E}^3$ is given by*

$$\eta: \mathbb{R}^3 \to \mathbb{R}^{(4,1)}, \quad P \mapsto o + p + \frac{1}{2}p^2 \infty.$$

1.5.2 Blades in CGA

Vectors and their geometric Interpretation Points $P \in \mathbb{R}^n$ are embedded into $\bigwedge^1 \mathbb{R}^{(n+1,1)}$ as null vectors. Moreover, it is clear that there are also other elements in $\bigwedge^1 \mathbb{R}^{(n+1,1)}$ that are not null vectors. Here we list these elements without discussing how to derive them. In the grade-1 space we have the following geometric entities, see [19]:

- Null vectors $\mathfrak{p} = \alpha(o + p + \frac{1}{2}p^2 \infty)$ representing a point.

- Vector without o-component: $\pi = n + \delta \infty$ representing a *dual hyperplane*. To determine the point set in $\mathbb{R}^n$ that is described by π we examine its inner product null space with respect to the description of points in CGA.

$$\mathbb{NI}_G(\pi) = \{\eta(x) \cdot \pi = 0 \mid x \in \mathbb{R}^3\} \tag{1.36}$$
$$= \left\{ (o + x + \frac{1}{2}x^2 \infty) \cdot (n + \delta \infty) = 0 \mid x \in \mathbb{R}^3 \right\}$$
$$= \{x \cdot n - \delta = 0 \mid x \in \mathbb{R}^3\}.$$

All $x \in \mathbb{NI}_G(\pi)$ given by Eq. (1.36) are contained by a hyperplane with normal vector n and distance $\frac{\delta}{\|n\|}$ to the origin. Note that ∞ is incident with every dual hyperplane.

- A general vector $s_\pm = \alpha(c \pm \frac{1}{2}\rho^2 \infty)$ defines a *dual hypersphere*. This can be interpreted as the sum of a point, *i.e.*, a null vector and an additional part $\sigma = \alpha(c + \beta \infty)$, where c is the normalized representation of a point. We can calculate the set of points contained in this object through

$$\mathbb{NI}_G(\sigma) = \{\eta(x) \cdot \sigma = 0 \mid x \in \mathbb{R}^3\} = \{\alpha(x \cdot c + \beta x \cdot \infty) = 0 \mid x \in \mathbb{R}^3\}$$
$$= \left\{ \alpha(-\frac{1}{2}\|x - c\|^2 - \beta) = 0 \mid x \in \mathbb{R}^3 \right\}.$$

Therefore, points contained in $\mathbb{NI}_G(\sigma)$ fullfill $\|x-c\|^2 = -2\beta$. If β is negative we redefine $\beta = -\frac{1}{2}\rho^2$. This results in the equation of a sphere with center c and radius ρ. Note that a positive value of β results in a sphere with real equation and imaginary radius. It carries no real point.

When speaking about dual objects we have to clarify what is meant by dualization. In CGA the dual of a blade is the orthogonal complement. For non-degenerate algebras this is usually done by multiplication with the pseudoscalar $\mathfrak{I} = o \wedge e_1 \wedge \ldots \wedge e_n \wedge \infty$, in the degenerate case we have to use the Poincaré duality cf. Def. 1.16.

Rounds and Flats In the homogeneous model lines can be defined by the exterior product of two points. Furthermore, planes can be defined by the exterior product of three non-collinear points and so on. In CGA a point is described by a null vector. The outer product of two points corresponds to a line in the homogeneous model that now has to be intersected with the horosphere. The result is the simplest round, *i.e.*, a pair of points. Let us take a look on the exterior product of three null vectors corresponding to three non-collinear points $\mathfrak{a} = \eta(a)$, $\mathfrak{b} = \eta(b)$ and $\mathfrak{c} = \eta(c)$, $a, b, c \in \mathbb{R}^3$:

$$\mathfrak{A} = \mathfrak{a} \wedge \mathfrak{b} \wedge \mathfrak{c}.$$

The intersection of this three-blade with the horosphere is an ellipse and the projection to $\mathbb{R}^n$ is a circle, except if one of the points is ∞. In the latter case the image of the ellipse is a line. Note that the conformal model is produced by compactifying the Euclidean space, *i.e.*, the affine real space with one point at infinity. Therefore, every flat has to contain this single point at infinity. To do kinematics we are interested in the three-dimensional Euclidean space modelled by the conformal geometric algebra $\mathcal{C}\ell_{(4,1,0)}$. Planes and spheres can be described by the exterior product of the null vectors corresponding to four non-concyclic respectively non-collinear points $a, b, c, d \in \mathbb{R}^3$. The corresponding null vectors are given by $\mathfrak{a} = \eta(a)$, $\mathfrak{b} = \eta(b)$, $\mathfrak{c} = \eta(c)$, $\mathfrak{d} = \eta(d)$ and the exterior product results in

$$\mathfrak{A} = \mathfrak{a} \wedge \mathfrak{b} \wedge \mathfrak{c} \wedge \mathfrak{d}.$$

1.5 The Conformal Model

In the homogeneous model this results in a three-space that has to be intersected with the horosphere and afterwards projected to $\mathbb{R}^3$ spanned by $[e_1, e_2, e_3]$. The result is a sphere if non of the points $\mathfrak{a}, \mathfrak{b}, \mathfrak{c}$, or $\mathfrak{d}$ is ∞ and a plane if one is ∞. All these objects can be understood as geometric outer product null spaces of blades. Their dual algebra elements $\mathfrak{A}\mathfrak{I}$ can be understood as geometric inner product null spaces of blades, where $\mathfrak{I}$ denotes the pseudoscalar.

Dual Objects Planes and spheres are modelled by decomposable elements of grade four. Dualization means multiplication with the pseudoscalar. Therefore, the dual elements are of grade one. Since the dual element describes the orthogonal complement of the object, it is clear why we compute $\mathfrak{A}\mathfrak{I} \cdot \eta x = 0$, $x \in \mathbb{R}^3$ to get the point set described by $\mathbb{NO}_G(\mathfrak{A}) = \mathbb{NI}_G(\mathfrak{A}\mathfrak{I})$.

Tangent Blades There are also linear subspaces in the homogeneous model that are tangent to the horosphere. It is clear that they cannot represent a point, but they should be interpreted as set of points, since they touch the horosphere in a specific null vector corresponding to a point. Furthermore, these objects posses another component that may be understood as a direction. One simple example of such a blade is constructed by the outer product of the origin with an arbitrary blade of the base space

$$\mathfrak{T} = o \wedge \mathfrak{a}.$$

The origin is the only intersection with the horosphere. Therefore, we think of $\mathfrak{a}$ as a direction appended to the origin. That is the reason why we call such blades *tangent blades*.

Free Blades A prototype of a *free blade* has the form

$$\mathfrak{F} = \mathfrak{a} \wedge \infty.$$

Note that it has no o-component. Such a blade is interpreted as a flat at infinity. Furthermore, translation has no effect on this blade, rotation just effects the $\mathfrak{a}$-part. Therefore, we interpret these blades as free direction vectors.

1.5.3 Conformal Transformations

The Pin group of $\mathcal{C}\ell_{(n+1,1,0)}$

$$\text{Pin}_{(n+1,1,0)} := \left\{ \mathfrak{g} \in \mathcal{C}\ell_{(n+1,1,0)} \,|\, N(\mathfrak{g}) = \pm 1,\, \alpha(\mathfrak{g})\mathfrak{v}\mathfrak{g}^* \in \bigwedge^1 V \,\,\forall \mathfrak{v} \in \bigwedge^1 V \right\}$$

is a double cover of the group of *Lorentz transformations*, *i.e.*, the group of orthogonal transformations of the Minkowski space $\mathbb{R}^{(n+1,1)}$. Furthermore, this group is isomorphic to the conformal group of $\mathbb{R}^n$, see [55]. The Spin group

$$\text{Spin}_{(n+1,1,0)} := \left\{ \mathfrak{g} \in \mathcal{C}\ell^+_{(n+1,1,0)} \,|\, N(\mathfrak{g}) = \pm 1,\, \mathfrak{g}\mathfrak{v}\mathfrak{g}^* \in \bigwedge^1 V \,\,\forall \mathfrak{v} \in \bigwedge^1 V \right\}$$

is a double cover of the orientation preserving conformal transformations. We construct this group with its subgroups by studying the grade-1 elements, that generate this group. First we identify the group SE(3) of Euclidean displacements as a subgroup of the conformal group. Therefore, we have to find all non-null vectors that leave the point at infinity fixed. This results in the condition

$$\alpha(\mathfrak{g}) \infty \mathfrak{g}^{-1} = \infty.$$

We derive $\infty \mathfrak{g} - \alpha(\mathfrak{g})\infty = 0$. The most general vector satisfying this condition is $\pi = n + \delta \infty$, see [19]. The geometric inner product null space corresponding to π is a plane.

Reflection in a Plane The most general versor fixing the point at infinity corresponds to a plane in inner product null space representation, *i.e.*, a dual plane. We study the action of this versor to a point. Furthermore, we are free to chose the origin. Therefore, we choose a plane passing through the origin, this means $\delta = 0$. We calculate

$$-n(o + p + \frac{1}{2}p^2 \infty)n^{-1} = o - npn^{-1} + \frac{1}{2}p^2 \infty$$

and see that this transformation leaves the point at infinity and the origin fixed. Furthermore, this transformation is a reflection in the plane defined by n. Two reflections in parallel planes result in a

1.5 The Conformal Model

translation written in versor form by $\mathfrak{t} = 1 - \frac{1}{2}t\infty$, $t = t_1 e_1 + t_2 e_2 + t_3 e_3 \in \mathbb{R}^3$. The composition of two reflections in planes passing through the origin yields a general rotation $\mathfrak{r} = \cos\frac{1}{2}\varphi - \sin(\frac{1}{2}\varphi)\mathfrak{R}$, where $\mathfrak{R}$ denotes a bivector corresponding to Plücker coordinates of the rotation axis, see Ex. 1.6. The action on a point P is computed by

$$\mathfrak{r}(o + p + \frac{1}{2}p^2\infty)\mathfrak{r}^{-1} = o + \mathfrak{r}p\mathfrak{r}^{-1} + \frac{1}{2}p^2\infty.$$

Hence, the generators o and ∞ are invariant under this transformation and the point P undergoes a rotation. It is clear that a general Euclidean displacement can be obtained by composing translations and rotations.

Reflection in a Sphere Grade-1 elements and their geometric inner product null spaces can be used to describe spheres. We do not discuss this topic in detail here. A full classification of conformal mappings and their generation can be found in [19]. The action of a versor corresponding to a dual sphere applied to a point is, in general, a reflection in this sphere. A scaling is the composition of two reflections in spheres with the same center and different radii (both real). A versor corresponding to a transformation that scales with respect to the origin by the factor e^γ is defined by:

$$\mathfrak{s} = \cosh\frac{\gamma}{2} + \sin\frac{\gamma}{2}(o \wedge \infty) = e^{\frac{1}{2}\gamma o \wedge \infty}.$$

By composing positive scalings and displacements the group of affine transformations can be described as subgroup of $\mathrm{Spin}_{(4,1,0)}$.

Transversions A transversion is the composition of an inversion in the unit sphere, a translation and a further inversion in the unit sphere. This transformation can be written in exponential form by

$$(o - \frac{1}{2}\infty)(1 - \frac{1}{2}t\infty)(0 - \frac{1}{2}\infty) = e^{\frac{1}{2}ot}.$$

Table 1.2 summarizes the basic operations in the conformal Clifford algebra model and their versors, see [19, p. 476].

Table 1.2: Basic transformations in the conformal model

Type of Operation	Explicit Form	Exp. Form
refl. in origin plane	n	none
refl. in real unit sphere	$c - \frac{1}{2}\infty$	none
refl. in origin	$o \wedge \infty$	none
rotation over φ in $\mathfrak{R}$-plane	$\cos(\frac{1}{2}\varphi) - \sin(\frac{1}{2}\varphi)\mathfrak{R}$	$e^{-\frac{1}{2}\varphi\mathfrak{R}}$
translation over t	$1 - \frac{1}{2}t\infty$	$e^{-\frac{1}{2}t \wedge \infty}$
scaling by e^γ	$\cosh(\frac{1}{2}\gamma) - \sinh(\frac{1}{2}\gamma)o \wedge \infty$	$e^{\frac{1}{2}\gamma o \wedge \infty}$
transversion over t	$1 + \frac{1}{2}ot$	$e^{\frac{1}{2}o \wedge t}$

1.6 A Clifford algebraic Approach to Line Geometry

In this section we aim at a Clifford algebraic description of the Klein model, cf. section 1.6. Therefore, we recall results that are already published in [40] and present the Clifford algebra $\mathcal{C}\ell_{(3,3,0)}$ as homogeneous model for the line space. It is well-known, that projective automorphisms of Klein's quadric induce projective transformations of $\mathbb{P}^3(\mathbb{R})$ and vise versa, see [56]. This is the reason why we consider this homogeneous model. Hence, it is possible to describe projective transformations as the action of the versor group of $\mathcal{C}\ell_{(3,3,0)}$. Especially the action of a reflection in the vector space, i.e., of a grade-1 element induces a null polarity in $\mathbb{P}^3(\mathbb{R})$, as we will show.

Remark 1.13. *This construction works for any quadric in any projective space. In this and the following sections we are interested in quadrics that serve as point models for other geometries.*

To build up the homogeneous model we use the quadratic form of Klein's quadric M_2^4, that is given by

$$Q = \begin{pmatrix} O & I \\ I & O \end{pmatrix},$$

where O is the 3×3 zero matrix and I the 3×3 identity matrix. The matrix Q that is used here corresponds to the polarity of Klein's

1.6 A Clifford algebraic Approach to Line Geometry

quadric, since multiplication with real scalars has no effect, see Eq. (1.13). As underlying vector space for the Clifford algebra we take $\mathbb{R}^6$ as vector space model for $\mathbb{P}^5(\mathbb{R})$. The corresponding Clifford algebra has signature $(p, q, r) = (3, 3, 0)$ (cf. [47]) and is of dimension $2^6 = 64$. Lines of $\mathbb{P}^3(\mathbb{R})$ represented by Plücker coordinates, see Def. 1.4, correspond to null vectors in this algebra, *i.e.*, vectors that square to zero. A vector is given by

$$\mathfrak{v} = x_1 e_1 + x_2 e_2 + x_3 e_3 + x_4 e_4 + x_5 e_5 + x_6 e_6$$

and its square is computed by

$$\mathfrak{v}\mathfrak{v} = 2(x_1 x_4 + x_2 x_5 + x_3 x_6). \tag{1.37}$$

Eq. (1.37) evaluates to zero if, and only if, the Plücker condition (1.12) is fulfilled, *i.e.*, if the point $X = (x_1, \ldots, x_6)^T \mathbb{R} \in \mathbb{P}^5(\mathbb{R})$ is contained by M_2^4, and therefore, describes a line in $\mathbb{P}^3(\mathbb{R})$. The norm of a vector equals

$$\mathfrak{v}\mathfrak{v}^* = -2(x_1 x_4 + x_2 x_5 + x_3 x_6).$$

The polarity of the metric quadric is given by multiplication with the pseudoscalar $\mathfrak{J} := e_{123456}$. Duality between subspaces of $\mathbb{P}^5(\mathbb{R})$ induced by the polarity is expressed by multiplication with the pseudoscalar. Outer product null spaces can be used to describe point sets corresponding to algebra elements. Moreover, the dual geometric entity with respect to the polarity Q is obtained with the inner product null space

$$\mathbb{NI}(\mathfrak{A}) = \mathbb{NO}(\mathfrak{A}\mathfrak{J}).$$

1.6.1 Collineations and Correlations in the Image Space

The Plücker line coordintes of a line spanned by two different points

$$X = x\mathbb{R} = (x_0, x_1, x_2, x_3)^T \mathbb{R}, Y = y\mathbb{R} = (y_0, y_1, y_2, y_3)^T \mathbb{R} \in \mathbb{P}^3(\mathbb{R})$$

were defined in section 1.2.1 by

$$p_{01} = \begin{vmatrix} x_0 & x_1 \\ y_0 & y_1 \end{vmatrix}, p_{02} = \begin{vmatrix} x_0 & x_2 \\ y_0 & y_2 \end{vmatrix}, p_{03} = \begin{vmatrix} x_0 & x_3 \\ y_0 & y_3 \end{vmatrix},$$

$$p_{23} = \begin{vmatrix} x_2 & x_3 \\ y_2 & y_3 \end{vmatrix}, p_{31} = \begin{vmatrix} x_3 & x_1 \\ y_3 & y_1 \end{vmatrix}, p_{12} = \begin{vmatrix} x_1 & x_2 \\ y_1 & y_2 \end{vmatrix}.$$

Now we ask for projective automorphisms of Klein's quadric induced by collineations or correlations in $\mathbb{P}^3(\mathbb{R})$. First, we transfer projective transformations acting on $\mathbb{P}^3(\mathbb{R})$ to automorphic collineations of M_2^4. Let $C = (c_{kl}), k, l = 0, \ldots, 3$ be the matrix representation of a collineation. We apply this collineation to the points $X = x\mathbb{R}$, $Y = y\mathbb{R} \in \mathbb{P}^3(\mathbb{R})$ with $x = (x_0, x_1, x_2, x_3)^T$, $y = (y_0, y_1, y_2, y_3)^T$ and compute the Plücker coordinates of the line joining $x' = Cx$ and $y' = Cy$. The Plücker coordinates of the image line under this collineation are given by:

$$p'_{ij} = \begin{vmatrix} x'_i & x'_j \\ y'_i & y'_j \end{vmatrix} = x'_i y'_j - x'_j y'_i$$
$$= \left(\sum_k c_{ik} x_k\right)\left(\sum_l c_{jl} y_l\right) - \left(\sum_l c_{jl} x_l\right)\left(\sum_k c_{ik} y_k\right)$$
$$= \sum_{k,l} c_{ik} c_{jl} (x_k y_l - x_l y_k),$$

where (i, j) is one of $(0, 1), (0, 2), (0, 3), (2, 3), (3, 1)$ or $(1, 2)$, see [56, p. 139]. If we write the action of this transformation on the space of lines as matrix vector product we get a 6×6 matrix L containing the coefficients from the equations above

$$(p'_{01}, p'_{02}, p'_{03}, p'_{23}, p'_{31}, p'_{12})^T = L \cdot (p_{01}, p_{02}, p_{03}, p_{23}, p_{31}, p_{12})^T.$$

When we repeat this procedure for a correlation the columns of the matrix correspond to plane coordinates. Hence, we can compute the collineation in the image space in the same way, but in this case we have to calculate the Plücker coordinates of the image lines by the intersection of two planes instead of the connection of two points. We get

$$(p'_{01}, p'_{02}, p'_{03}, p'_{23}, p'_{31}, p'_{12})^T = \bar{L} \cdot (p_{01}, p_{02}, p_{03}, p_{23}, p_{31}, p_{12})^T, \quad (1.38)$$

where $\bar{L}$ defines an automorphic collineation of M_2^4 corresponding to a correlation in $\mathbb{P}^3(\mathbb{R})$.

Now we examine geometric entities that are described in this geometric algebra by inner product and outer product null spaces, see [53].

1.6.2 Algebra Representation of Linear Line Manifolds

In this section we introduce the Clifford algebra representation for the occurring linear line manifolds as outer product null spaces of $(k+1)$-blades. Using duality k-space intersections can also be described as inner product null space of $(n-(k+1))$-blades where $n=6$ is the dimension of the vector space model of $\mathbb{P}^5(\mathbb{R})$. We start with subspaces contained entirely in Klein's quadric.

Lines contained in M_2^4 as null two-blades Null blades can be used to describe subspaces contained entirely in M_2^4. Subspaces that are contained in M_2^4 are either lines or two-spaces in $\mathbb{P}^5(\mathbb{R})$. A null two-blade generated by the exterior product of two null vectors corresponding to conjugate points on M_2^4 defines a line in $M_2^4 \subset \mathbb{P}^5(\mathbb{R})$. Its outer product null space is the set of all null vectors corresponding to a pencil of lines in $\mathbb{P}^3(\mathbb{R})$.

Two-spaces contained in M_2^4 as null three-blades A two-space P_1^2 in $\mathbb{P}^5(\mathbb{R})$ that is contained entirely in Klein's quadric can be expressed as the exterior product of three null vectors corresponding to points contained in the two-space. This results in a null three-blade. Its outer product null space consists of all null vectors that correspond to points contained in the two-space $P_1^2 \subset M_2^4$, i.e., a bundle of lines or a field of lines.

Conics on Klein's quadric as non-null three-blades Three-blades corresponding to two-spaces in $\mathbb{P}^5(\mathbb{R})$ can be defined as exterior product of three null vectors $\mathfrak{v}_1, \mathfrak{v}_2, \mathfrak{v}_3 \in \bigwedge^1 V$ corresponding to points on M_2^4. If the three-blade $\mathfrak{B} \in \bigwedge^3 V$ squares to zero it corresponds to a two-space that is entirely contained in M_2^4, else it corresponds to a two-space that intersects in a conic on $M_2^4 \subset \mathbb{P}^5(\mathbb{R})$. All points contained by this two-space can be computed with the help of the outer product null space

$$\mathbb{NO}(\mathfrak{v}_1 \wedge \mathfrak{v}_2 \wedge \mathfrak{v}_3) = \mathbb{NO}(\mathfrak{v}_1) \oplus \mathbb{NO}(\mathfrak{v}_2) \oplus \mathbb{NO}(\mathfrak{v}_3).$$

To get the null vectors located in the two-space $\mathfrak{p} = \alpha \mathfrak{v}_1 + \beta \mathfrak{v}_2 + \gamma \mathfrak{v}_3$ we determine the zero divisors by $\mathfrak{p}\mathfrak{p} = 0$. This results in a quadratic equation involving the coefficients α, β, and γ. The solution is given by the intersection of Klein's quadric with the two-space. In $\mathbb{P}^5(\mathbb{R})$ the dual of a two-space is a two-space and the points contained by the dual of a two-space can be calculated by the inner product null space of a three-blade corresponding to the two-space.

Linear line congruences as four-blades Three-spaces are polar to lines. Thus, linear line congruences can be described by inner product null spaces of two-blades that correspond to lines in $\mathbb{P}^5(\mathbb{R})$ or outer product null spaces of four-blades that correspond to three-spaces in $\mathbb{P}^5(\mathbb{R})$. Lines in $\mathbb{P}^5(\mathbb{R})$ are represented by the exterior product of two vectors $\mathfrak{v}_1, \mathfrak{v}_2 \in \bigwedge^1 V$ corresponding to points in $\mathbb{P}^5(\mathbb{R})$. A general line in $\mathbb{P}^5(\mathbb{R})$ written as outer product of two arbitrary vectors

$$\mathfrak{v}_1 = x_1 e_1 + x_2 e_2 + x_3 e_3 + x_4 e_4 + x_5 e_5 + x_6 e_6,$$
$$\mathfrak{v}_2 = y_1 e_1 + y_2 e_2 + y_3 e_3 + y_4 e_4 + y_5 e_5 + y_6 e_6$$

has the form

$$\mathfrak{L} = \mathfrak{v}_1 \wedge \mathfrak{v}_2 = \sum_{\substack{i,j=1 \\ i<j}}^{6} \begin{vmatrix} x_i & x_j \\ y_i & y_j \end{vmatrix} e_{ij}.$$

With Eq. (1.30) we know that the two conditions

$$a_1 x_4 + a_2 x_5 + a_3 x_6 + a_4 x_1 + a_5 x_2 + a_6 x_3 = 0,$$
$$a_1 y_4 + a_2 y_5 + a_3 y_6 + a_4 y_1 + a_5 y_2 + a_6 y_3 = 0$$

are sufficient to describe the inner product null space of $\mathfrak{L}$. All null vectors $\mathfrak{a} = a_1 e_1 + a_2 e_2 + a_3 e_3 + a_4 e_4 + a_5 e_5 + a_6 e_6$ satisfying these conditions correspond to the set of lines contained in a linear line congruence. In a dual way, a linear line congruence can also be represented as the outer product null space of a four-blade that can be constructed as exterior product of four linearly independent vectors.

Linear Line Complexes as five-blades A four-space in $\mathbb{P}^5(\mathbb{R})$ can be described as the outer product null space of a five-blade $\mathfrak{B} \in \bigwedge^5 V$. Using duality, the same four-space is defined by the inner product null space of the vector $\mathfrak{B}\mathfrak{J} \in \bigwedge^1 V$. If the given vector is a null vector, the linear line complex is singular, else regular. The outer product null space of a vector and the inner product null space of its dual correspond to the vector itself. Let $\mathfrak{v} = x_1 e_1 + x_2 e_2 + x_3 e_3 + x_4 e_4 + x_5 e_5 + x_6 e_6$ be a general vector. Then its inner product null space is given by

$$\mathbb{NI}(\mathfrak{v}) = \left\{ \mathfrak{a} \in \bigwedge^1 V \mid x_1 a_4 + x_2 a_5 + x_3 a_6 + x_4 a_1 + x_5 a_2 + x_6 a_3 = 0 \right\},$$

with $\mathfrak{a} = a_1 e_1 + a_2 e_2 + a_3 e_3 + a_4 e_4 + a_5 e_5 + a_6 e_6$. The same set of lines can be obtained as the outer product null space of the dual of $\mathfrak{v}$:

$$\mathbb{NO}(\mathfrak{v}\mathfrak{J}) = \left\{ \mathfrak{a} \in \bigwedge^1 V \mid x_1 a_4 + x_2 a_5 + x_3 a_6 + x_4 a_1 + x_5 a_2 + x_6 a_3 = 0 \right\},$$

with $\mathfrak{a} = a_1 e_1 + a_2 e_2 + a_3 e_3 + a_4 e_4 + a_5 e_5 + a_6 e_6$. Note, that it is not important from which side we multiply with the pseudoscalar, since $\mathfrak{J}\mathfrak{v} = -\mathfrak{v}\mathfrak{J}$. We are working in a projective setting, and thus, the multiplication by -1 or by a real number does not change the geometric meaning of the object and its inner product or outer product null space.

1.6.3 Transformations

To describe the action of the versor group we recall the definition of a versor, see Def. 1.10. A *versor* $\mathfrak{a}$ is an algebra element that can be expressed as the k-fold geometric product of non-null vectors. Thus, $\mathfrak{a} = \mathfrak{v}_1 \ldots \mathfrak{v}_k$ with $\mathfrak{v}_i \in \bigwedge^1 V$, $i = 1, \ldots, k$. Transformations are applied with the so-called sandwich operator.

Remark 1.14. *For a vector $\mathfrak{v} \in \bigwedge^1 V$ the inverse element can be obtained by*
$$\mathfrak{v}^{-1} = \frac{\mathfrak{v}^*}{\mathfrak{v}\mathfrak{v}^*}.$$
Thus, the inverse element differs from the conjugate element by a real factor. This is also true for a k-fold product of vectors.

Since we are working in a homogeneous Clifford algebra model multiplication with a real factor does not change the geometric meaning of an algebra element. Therefore, we use the conjugate element instead of the inverse element. Hence, the sandwich operator that we use is given by:
$$\alpha(\mathfrak{a})\mathfrak{v}\mathfrak{a}^* = \mathfrak{a}\mathfrak{a}^*(\alpha(\mathfrak{a})\mathfrak{v}\mathfrak{a}^{-1}).$$
This operator does not involve an inverse, and therefore, it can also be applied if the element $\mathfrak{a}$ is not invertible.

If a vector $\mathfrak{v} \in \bigwedge^1 V$ corresponds to a line $L \in \mathcal{L}^3$ it is a null vector respectively a zero divisor, *i.e.*, it has no inverse. Applying the sandwich operator with an arbitrary element of the versor group to a zero divisor results again in a zero divisor since the zero divisors forms an ideal in the algebra. This means null vectors corresponding to lines in $\mathbb{P}^3(\mathbb{R})$ are mapped to null vectors corresponding to lines in $\mathbb{P}^3(\mathbb{R})$. The transformations induced by non-null vectors are reflections in $\mathbb{R}^6$ as a model for $\mathbb{P}^5(\mathbb{R})$ that fix Klein's quadric.

The advantage of this model is that transformations can be applied via the sandwich operator to every entity that can be represented as $\mathfrak{A} \in \bigwedge^k V$ in inner product or outer product null space representation. For example, we are able to apply the transformation to a four-blade that corresponds to a linear line congruence with one sandwich operator. The resulting element is again a four-blade corresponding to a linear line congruence that is of the same projective type (elliptic, parabolic, or hyperbolic) as the original one. At this point we recall a theorem from [56, theorem 2.1.10].

Theorem 1.2. *Projective collineations and correlations of $\mathbb{P}^3(\mathbb{R})$ induce projective automorphisms of Klein's quadric, and Klein's quadric does not admit other projective automorphisms.*

1.6 A Clifford algebraic Approach to Line Geometry

The Clifford group of this Clifford algebra, *i.e.*, the group of projective automorphisms of Klein's quadric is generated by non-null vectors. Thus, we are interested in the action of non-null vectors on null vectors. Let

$$\mathfrak{a} = a_1 e_1 + a_2 e_2 + a_3 e_3 + a_4 e_4 + a_5 e_5 + a_6 e_6,$$
$$\mathfrak{v} = x_1 e_1 + x_2 e_2 + x_3 e_3 + x_4 e_4 + x_5 e_5 + x_6 e_6$$

be two vectors with

$$\mathfrak{a}\mathfrak{a} = a_1 a_4 + a_2 a_5 + a_3 a_6 \neq 0, \qquad \mathfrak{v}\mathfrak{v} = x_1 x_4 + x_2 x_5 + x_3 x_6 = 0.$$

The action of the sandwich operator $\alpha(\mathfrak{a})\mathfrak{v}\mathfrak{a}^*$ is linear on the vector space $\bigwedge^1 V$. The matrix acting on $\mathbb{P}^5(\mathbb{R})$ can be represented by

$$M = \begin{pmatrix} k_1 & a_1 a_5 & a_1 a_6 & a_1 a_1 & a_1 a_2 & a_1 a_3 \\ a_2 a_4 & k_2 & a_2 a_6 & a_2 a_1 & a_2 a_2 & a_2 a_3 \\ a_3 a_4 & a_3 a_5 & k_3 & a_3 a_1 & a_3 a_2 & a_3 a_3 \\ a_4 a_4 & a_4 a_5 & a_4 a_6 & k_4 & a_4 a_2 & a_4 a_3 \\ a_5 a_4 & a_5 a_5 & a_5 a_6 & a_5 a_1 & k_5 & a_5 a_3 \\ a_6 a_4 & a_6 a_5 & a_6 a_6 & a_6 a_1 & a_6 a_2 & k_6 \end{pmatrix}, \qquad (1.39)$$

with

$$k_1 = -a_5 a_2 - a_6 a_3, \qquad k_2 = -a_6 a_3 - a_4 a_1, \qquad k_3 = -a_4 a_1 - a_5 a_2,$$
$$k_4 = -a_5 a_2 - a_6 a_3, \qquad k_5 = -a_6 a_3 - a_4 a_1, \qquad k_6 = -a_4 a_1 - a_5 a_2.$$

Naturally, we now ask for the corresponding projective mapping acting on $\mathbb{P}^3(\mathbb{R})$. Therefore, we examine the action of the collineation M on the space of lines. The action on the base lines corresponding to the Plücker coordinate vectors

$$b_1 = (1:0:0:0:0:0), \qquad b_2 = (0:1:0:0:0:0), \qquad b_3 = (0:0:1:0:0:0),$$
$$b_4 = (0:0:0:1:0:0), \qquad b_5 = (0:0:0:0:1:0), \qquad b_6 = (0:0:0:0:0:1)$$

is given by the columns of the matrix M.

$$h_1 = Mb_1, \quad h_2 = Mb_2, \quad h_3 = Mb_3, \quad h_4 = Mb_4, \quad h_5 = Mb_5, \quad h_6 = Mb_6.$$

If we look at the images of the three ideal lines b_4, b_5, b_6 we see, that the corresponding lines posses linear dependent direction vectors. This means that the three image lines h_4, h_5, h_6 intersect in the same ideal point, and therefore, they belong to a bundle of lines. Since the lines b_4, b_5, b_6 define a field of lines and the image is a bundle of lines, the mapping must be a correlation. With this knowledge we determine the action of M on the bundle of lines concurrent to the origin spanned by b_1, b_2, b_3. The image of this bundle of lines is a field of lines. To obtain the coordinates of the plane that carries the field of lines we intersect the image lines and get three points defining the plane in $\mathbb{P}^3(\mathbb{R})$

$$h_1 \cap h_2 = s_1 = (a_3, -a_5, a_4, 0)^\mathrm{T},$$
$$h_1 \cap h_3 = s_2 = (a_2, a_6, 0, -a_4)^\mathrm{T},$$
$$h_2 \cap h_3 = s_3 = (a_1, 0, -a_6, a_5)^\mathrm{T}.$$

The plane coordinates p_1 of the plane generated by these three points have to satisfy

$$\langle p_1, s_1 \rangle = \langle p_1, s_2 \rangle = \langle p_1, s_3 \rangle = 0. \tag{1.40}$$

Solving Eq. 1.40 results in $p_1 = \mathbb{R}(0, a_4, a_5, a_6)^\mathrm{T}$ as image of the origin $O = (1, 0, 0, 0)^\mathrm{T}\mathbb{R}$. We repeat this procedure for the bundle of lines spanned by b_1, b_5, b_6, i.e., the bundle of lines concurrent to the ideal point $X_u = (0, 1, 0, 0)^\mathrm{T}\mathbb{R}$.

$$h_1 \cap h_5 = (a_2, a_6, 0, -a_4)^\mathrm{T}, \qquad h_1 \cap h_6 = (a_3, -a_5, a_4, 0)^\mathrm{T},$$
$$h_5 \cap h_6 = (0, a_1, a_2, a_3)^\mathrm{T}, \qquad p_2 = \mathbb{R}(-a_4, 0, a_3, -a_2)^\mathrm{T}.$$

Repeating the procedure for the bundle of lines concurrent to the ideal point $Y_u = (0, 0, 1, 0)^\mathrm{T}\mathbb{R}$ spanned by the lines b_2, b_4, and b_6 yields

$$h_2 \cap h_4 = (a_1, 0, -a_6, a_5)^\mathrm{T}, \qquad h_2 \cap h_6 = (a_3, -a_5, a_4, 0)^\mathrm{T},$$
$$h_4 \cap h_6 = (0, a_1, a_2, a_3)^\mathrm{T}, \qquad p_3 = \mathbb{R}(-a_5, -a_3, 0, a_1)^\mathrm{T}.$$

Analogue computation for the bundle of lines spanned by b_3, b_4, and b_5 concurrent to the ideal point $Z_u = (0, 0, 0, 1)^\mathrm{T}\mathbb{R}$ results in

$$h_3 \cap h_4 = (a_1, 0, -a_6, a_5)^\mathrm{T}, \qquad h_3 \cap h_5 = (a_2, a_6, 0, -a_4)^\mathrm{T},$$
$$h_4 \cap h_5 = (0, a_1, a_2, a_3)^\mathrm{T}, \qquad p_4 = \mathbb{R}(-a_6, a_2, -a_1, 0)^\mathrm{T}.$$

Note, that the correlation is not determined through $p_1, \ldots, p_4$ since the scaling of each p_i, $i = 1, \ldots, 4$ is not determined yet. Together

1.6 A Clifford algebraic Approach to Line Geometry

with the matrix representation of a correlation see Eq. (1.38) the correlation in $\mathbb{P}^3(\mathbb{R})$ is fixed. It can be written in matrix form as

$$K = \begin{pmatrix} 0 & -a_4 & -a_5 & -a_6 \\ a_4 & 0 & -a_3 & a_2 \\ a_5 & a_3 & 0 & -a_1 \\ a_6 & -a_2 & a_1 & 0 \end{pmatrix}.$$

This correlation is a null polarity, see [65]. The determinant of this null polarity is calculated by

$$\det K = (a_1 a_4 + a_2 a_5 + a_3 a_6)^2 = \frac{1}{4}(\mathfrak{aa})^2. \tag{1.41}$$

Thus, the square of a vector is related to the determinant of the corresponding null polarity.

Remark 1.15. *Note, that this procedure works also for null vectors, because we do not use the inverse element to determine the sandwich product in homogeneous Clifford Algebra models.*

With these preliminaries we are now able to prove the following theorem:

Theorem 1.3. *Each regular projective transformation, i.e., regular correlation or regular colline-ation can be represented by the product of at most six null polarities.*

Proof. The group of automorphic collineations of M_2^4 and projective transformations of $\mathbb{P}^3(\mathbb{R})$ are isomorphic, see [56]. Furthermore, the group of automorphisms of Klein's quadric can be described by the Pin group of $\mathcal{Cl}_{(3,3,0)}$, see [28] for the general case. The Clifford algebra model has at most grade six. Furthermore, the k-fold product of grade-1 elements, that correspond to null polarities generate the Pin or Spin group depending on whether k is odd or even. To reach the maximum grade of six, we need a product of at least six grade-1 elements. That at most six elements are necessary follows from the Cartan-Dieudonné theorem, cf. [22]. □

All together we have that the versor group generated by reflections respectively non-null vectors corresponds to the group of automorphic collineations of M_2^4 induced by regular projective transformations in $\mathbb{P}^3(\mathbb{R})$ that can be written as the product of invulotic automorphic collineations of Klein's quadric corresponding to null polarities. Furthermore, all transformations that are generated by the product of an even number of vectors are collineations, *i.e.*, elements of the Spin group when they are normalized. Transformations that are generated by an odd number of vectors correspond to correlations, and therefore, to the Pin group when they are normalized.

Example 1.8. *As example we present the action of a null polarity applied to a conic on Klein's quadric. Thus, we give three null vectors corresponding to Plücker coordinates l_i, $i = 1, 2, 3$ of three lines*

$$\mathfrak{l}_1 = e_1 + 2e_6, \quad \mathfrak{l}_2 = e_2 + 2e_4, \quad \mathfrak{l}_3 = e_3 + 2e_5.$$

The corresponding two-space in $\mathbb{P}^5(R)$ is represented as three-blade by

$$\mathfrak{P} = \mathfrak{l}_1 \wedge \mathfrak{l}_2 \wedge \mathfrak{l}_3 = e_{123} + 2e_{236} - 2e_{134} - 4e_{346} + 2e_{125} + 4e_{256} + 4e_{145} + 8e_{456}.$$

The outer product null space of this entity can be calculated as

$$\mathbb{NO}(\mathfrak{P}) = \left\{ \alpha \left(\frac{1}{2}e_2 + e_4 \right) + \beta \left(\frac{1}{2}e_3 + e_5 \right) + \gamma \left(\frac{1}{2}e_1 + e_6 \right) \mid \alpha, \beta, \gamma \in \mathbb{R} \right\}.$$

Indeed, the outer product null space of $\mathfrak{P}$ spans the same two-space as $[l_1, l_2, l_3]$. To get the regulus defined by the three lines L_i or their corresponding null vectors $\mathfrak{l}_i$ we have to find all null vectors in the two-space $\mathfrak{p} = \alpha\mathfrak{l}_1 + \beta\mathfrak{l}_2 + \gamma\mathfrak{l}_3$. Therefore, we compute

$$\mathfrak{p}^2 = 4\beta\alpha + 4\gamma\alpha + 4\gamma\beta \stackrel{!}{=} 0. \tag{1.42}$$

The parameters $(\alpha : \beta : \gamma)$ can be interpreted as homogeneous coordinates in a projective plane P_1^2. Eq. (1.42) defines a conic contained in this plane. Now we apply an arbitrary null polarity given by

$$\mathfrak{a} = a_1 e_1 + a_2 e_2 + a_3 e_3 + a_4 e_4 + a_5 e_5 + a_6 e_6.$$

The outer product null space of the resulting three-blade $\mathfrak{P}' = \alpha(\mathfrak{a})\mathfrak{P}\mathfrak{a}^*$ can be calculated as the set

$$\mathbb{NO}(\mathfrak{P}') = \{\alpha\mathfrak{v}_1 + \beta\mathfrak{v}_2 + \gamma\mathfrak{v}_3 \mid \alpha, \beta, \gamma \in \mathbb{R}\},$$

1.6 A Clifford algebraic Approach to Line Geometry

where the vectors $\mathfrak{v}_i$ are given by

$$\mathfrak{v}_1 = \frac{2a_5a_1 - a_6a_5 - 2a_6a_1 + 4a_1^2}{a_5a_4 + a_6a_4 + a_6a_5}e_1 + \frac{a_6a_5 + 2a_5a_2 + 4a_1a_2 - 2a_4a_1 + a_6a_4}{a_5a_4 + a_6a_4 + a_6a_5}e_2$$
$$+ \frac{2a_5a_3 - a_5^2 + 4a_1a_3 - 2a_5a_1}{a_5a_4 + a_6a_4 + a_6a_5} + 2e_4,$$

$$\mathfrak{v}_2 = \frac{4a_1a_2 + 2a_6a_1 - a_6^2 - 2a_6a_2}{a_5a_4 + a_6a_4 + a_6a_5}e_1 + \frac{2a_6a_2 - 2a_4a_2 - a_6a_4 + 4a_2^2}{a_5a_4 + a_6a_4 + a_6a_5}e_2$$
$$+ \frac{a_5a_4 + 2a_6a_3 + a_6a_4 + 4a_3a_2 - 2a_5a_2}{a_5a_4 + a_6a_4 + a_6a_5}e_3 + 2e_5,$$

$$\mathfrak{v}_3 = \frac{2a_4a_1 + 4a_1a_3 + a_6a_5 - 2a_6a_3 + a_5a_4}{a_5a_4 + a_6a_4 + a_6a_5}e_1 + \frac{2a_4a_2 - 2a_4a_3 + 4a_3a_2 - a_4^2}{a_5a_4 + a_6a_4 + a_6a_5}e_2$$
$$+ \frac{2a_4a_3 - a_5a_4 + 4a_3^2 - 2a_5a_3}{a_5a_4 + a_6a_4 + a_6a_5}e_3 + 2e_6.$$

We search for null vectors contained in the plane $\mathfrak{p}'(\alpha, \beta, \gamma) = \alpha\mathfrak{v}_1 + \beta\mathfrak{v}_2 + \gamma\mathfrak{v}_3$. Therefore, we calculate $\mathfrak{p}'\mathfrak{p}' = 0$. This leads to the following quadratic equation

$$((-2a_4a_3 + a_5a_4 + 2a_6a_3 + a_6a_4 + 2a_4a_2 + 8a_3a_2 - 2a_5a_2 - a_4^2)\beta$$
$$+ (8a_1a_3 - 2a_6a_3 - 2a_5a_1 + 2a_4a_1 + 2a_5a_3 + a_6a_5 - a_5^2 + a_5a_4)\alpha)\gamma$$
$$+ (-2a_4a_1 + 2a_5a_2 + 2a_6a_1 + a_6a_4 + a_6a_5 + 8a_1a_2 - 2a_6a_2 - a_6^2)\alpha\beta$$
$$+ (2a_5a_1 - a_6a_5 - 2a_6a_1 + 4a_1^2) + (2a_4a_3 - a_5a_4 + 4a_3^2 - 2a_5a_3)\gamma^2\alpha^2$$
$$+ (2a_6a_2 - 2a_4a_2 - a_6a_4 + 4a_2^2)\beta^2 = 0.$$

This equation defines a conic in the image of the projective plane P_1^2 under the null polarity induced by $\mathfrak{a}$. Every point on this conic delivers the parameters for a null vector that corresponds to a line in $\mathbb{P}^3(\mathbb{R})$.

1.6.4 Collineations as Spin Group

We are interested in the relationship between regular projective mappings and elements of the versor group of $\mathcal{C}\ell_{(3,3,0)}$. The general approach (cf.[53]) does not work for this model. Therefore, we develop a more line geometric approach by the examination of the action of versors on null three-blades corresponding to two-spaces that are entirely contained in M_2^4. A general element $\mathfrak{g} \in \mathcal{C}\ell_{(3,3,0)}^+$ corresponding to a collineation is given by

$$\mathfrak{g} = g_1 e_0 + g_2 e_{12} + g_3 e_{13} + g_4 e_{14} + g_5 e_{15} + g_6 e_{16} + g_7 e_{23} + g_8 e_{24} + g_9 e_{25} + g_{10} e_{26}$$
$$+ g_{11} e_{34} + g_{12} e_{35} + g_{13} e_{36} + g_{14} e_{45} + g_{15} e_{46} + g_{16} e_{56} + g_{17} e_{1234} + g_{18} e_{1235}$$
$$+ g_{19} e_{1236} + g_{20} e_{1245} + g_{21} e_{1246} + g_{22} e_{1256} + g_{23} e_{1345} + g_{24} e_{1346} + g_{25} e_{1356}$$
$$+ g_{26} e_{1456} + g_{27} e_{2345} + g_{28} e_{2346} + g_{29} e_{2356} + g_{30} e_{2456} + g_{31} e_{3456} + g_{32} e_{123456}.$$

The conditions that this element is the product of invertible vectors is obtained by

$$\alpha(\mathfrak{g}) \mathfrak{v} \mathfrak{g}^* \in \bigwedge\nolimits^1 V \text{ for all } \mathfrak{v} \in \bigwedge\nolimits^1 V. \tag{1.43}$$

This results in 36 quadratic equations that occur as coefficients of the grade-5 element e_J, where $J \subset \{1, 2, 3, 4, 5, 6\}$ with $|J| = 5$. The 36 quadratic equations defined above define a pseudo algebraic variety in $\mathbb{P}^{31}(\mathbb{R})$ that can be interpreted as image space of $\text{Spin}_{(3,3)}$, see chapter 3. Points of $\mathbb{P}^3(\mathbb{R})$ can be embedded in the homogeneous Clifford algebra as null three-blades that correspond to two-spaces on Klein's quadric. These two-spaces correspond to bundles of lines, *i.e.*, all lines concurrent to a point. The point contained by all lines of a bundle can be determined and is described uniquely by the null three-blade. Therefore, we examine the action of an arbitrary versor corresponding to a collineation on a null three-blade corresponding to a bundle of lines. This results in a null three-blade that corresponds to a bundle of lines again. Afterwards we compute the point concurrent to all lines contained by the bundle of lines to get the image of the start point under the transformation.

We start with the bundle of lines concurrent to the a general point $P = (y_0, y_1, y_2, y_3)^T \mathbb{R}$. The Plücker coordinates of three lines containing P are computed as

$$l_1 = (y_0 : 0 : 0 : 0 : y_3 : -y_2), \quad l_2 = (0 : y_0 : 0 : -y_3 : 0 : y_1), \quad l_3 = (0 : 0 : y_0 : y_2 : -y_1 : 0).$$

The null three-blade corresponding to the two-space spanned by l_1, l_2, and l_3 is obtained by

$$\mathfrak{b} = (y_0 e_1 + y_3 e_5 - y_2 e_6) \wedge (y_0 e_2 - y_3 e_4 + y_1 e_6) \wedge (y_0 e_3 + y_2 e_4 - y_1 e_5)$$

1.6 A Clifford algebraic Approach to Line Geometry

$$= y_0^3 e_{123} + y_3 y_0^2 e_{235} - y_2 y_0^2 e_{236} + y_3 y_0^2 e_{134} + y_3^2 y_0 e_{345} - y_2 y_3 y_0 e_{346}$$
$$- y_0^2 y_1 e_{136} + y_3 y_1 y_0 e_{356} + y_2 y_0^2 e_{124} + y_2 y_3 y_0 e_{245} - y_2^2 y_0 e_{246}$$
$$- y_0 y_1 y_2 e_{146} - y_0^2 y_1 e_{125} + y_0 y_1 y_2 e_{256} + y_3 y_1 y_0 e_{145} + y_0 y_1^2 e_{156}.$$

We apply the sandwich operator with the arbitrary element $\mathfrak{g} \in C\ell_{(3,3,0)}^+$. The outer product null space of the null three-blade $\mathfrak{b}' = \alpha(\mathfrak{g})\mathfrak{b}\mathfrak{g}^*$ is obtained by

$$\mathbb{NO}(\mathfrak{b}') = \{\alpha\mathfrak{v}_1 + \beta\mathfrak{v}_2 + \gamma\mathfrak{v}_3 | \alpha, \beta, \gamma \in \mathbb{R}\},$$

where

$$\mathfrak{v}_1 = \bigl(y_0(g_1 - g_{32} - g_{20} + g_4 - g_{29} + g_{13} + g_9 - g_{24})$$
$$+ 2y_1(g_{17} + g_7) + 2y_2(g_{18} - 2g_3) + 2y_3(g_2 + 2g_{19})\bigr)e_1$$
$$- \bigl(2y_0(g_{31} + g_{14}) + 2y_1(g_{27} - g_{11}) - 2y_2(g_{12}y_2 + g_{23})$$
$$+ y_3(g_9 - g_1 + g_{32} - g_{29} + g_4 - g_{13} + g_{20} - g_{24})\bigr)e_5$$
$$- \bigl(+ 2y_0(g_{15} - g_{30}) + 2y_1(g_8 + g_{28}) + 2y_3(g_{10} + g_{21})$$
$$+ y_2(g_9 - g_{24} - g_4 + g_1 - g_{32} + g_{20} + g_{29} - g_{13})\bigr)e_6,$$
$$\mathfrak{v}_2 = \bigl(y_0(g_1 - g_{29} + g_{13} + g_9 - g_{24} - g_{32} - g_{20} + g_4)$$
$$+ 2y_1(g_{17} + g_7) + 2y_2(g_{18} - g_3) + 2y_3(g_{19} + g_2)\bigr)e_2$$
$$+ \bigl(2y_0(g_{14} + g_{31}) + 2y_1(g_{27} - g_{11}) - 2y_2(g_{12}y_2 + g_{23})$$
$$+ y_3(g_9 - g_1 + g_{32} - g_{29} + g_4 - g_{13} + g_{20} - g_{24})\bigr)e_4$$
$$- \bigl(2y_0(g_{16} + g_{26}) - 2y_2(g_5 + g_{25}) + 2y_3(g_{22} - g_6)$$
$$+ y_1(g_{13} + g_{32} - g_4 + g_{29} + g_9 - g_{24} - g_{20} - g_1)\bigr)e_6,$$
$$\mathfrak{v}_3 = \bigl(y_0(g_1 - g_{29} + g_{13} + g_9 - g_{24} - g_{32} - g_{20} + g_4)$$
$$+ 2y_1(g_{17} + g_7) + 2y_2(g_{18} - g_3) + 2y_3(g_2 + g_{19})\bigr)e_3$$
$$+ \bigl(2y_0(g_{15} - g_{30}) + 2y_1(g_8 + g_{28}) + 2y_3(g_{10} + g_{21})$$
$$+ y_2(g_1 - g_{24} - g_4 - g_{32} + g_9 + g_{20} + g_{29} - g_{13})\bigr)e_4$$
$$+ \bigl(2y_0(g_{16} + g_{26}) + 2y_3(g_{22} - g_6) - 2y_2(g_5 + g_{25})$$
$$+ y_1(g_{13} + g_{29} + g_{32} - g_4 + g_9 - g_{24} - g_{20} - g_1)\bigr)e_5.$$

The null vectors $\mathfrak{v}_1, \mathfrak{v}_2$, and $\mathfrak{v}_3$ span a bundle of lines concurrent to the image of the origin under the collineation corresponding to $\mathfrak{g}$. To get the point we have to intersect two lines of the bundle of lines. Therefore, we change the model and transfer the null vectors corresponding to $\mathfrak{v}_1$, $\mathfrak{v}_2$ to two-blades $\mathfrak{L}_1$, $\mathfrak{L}_2$ of the Grassmann algebra $\mathcal{G}(\mathbb{P}^3)$, cf. [27]. The representation of a line $L = (p_{01}:p_{02}:p_{03}:p_{23}:p_{31}:p_{12})$ in $\mathcal{G}(\mathbb{P}^3)$ is

$$L = p_{01}e_{12} + p_{02}e_{13} + p_{03}e_{14} + p_{23}e_{34} - p_{31}e_{24} + p_{12}e_{23}. \qquad (1.44)$$

Note that due to historic conversion the coefficient of e_{24} is multiplied by -1. We obtain

$$\begin{aligned}
\mathfrak{L}_1 =\ & (y_0(g_1 + g_4 + g_9 - g_{29} + g_{13} - g_{20} - g_{32} - g_{24}) + 2y_1(g_{17} + g_7) \\
& + 2y_2(g_{18} - g_3) + 2y_3(g_2 + g_{19}))e_{12} \\
& + (2y_0(g_{31} + g_{14}) + 2(g_{27} - g_{11}) - 2y_2(g_{12} + g_{23}) \\
& + y_3(g_{32} + g_{20} + g_9 - g_1 - g_{13} - g_{29} + g_4 - g_{24}))e_{24} \\
& + (2y_0(g_{30} - g_{15}) - 2y_1(g_8 + g_{28}) \\
& + y_2(g_{24} + g_4 - g_1 - g_9 - g_{20} - g_{29} + g_{32} + g_{13}) - 2y_3(g_{10} + g_{21}))e_{23}, \\
\mathfrak{L}_2 =\ & (y_0(g_4 + g_9 - g_{29} + g_{13} - g_{20} + g_1 - g_{32} - g_{24}) + 2y_1(g_{17} + g_7) \\
& + 2y_2(g_{18} - g_3) + 2y_3(g_2 + g_{19}))e_{13} \\
& + (2y_0(g_{14} + g_{31}) + 2y_1(g_{27} - g_{11}) - 2y_2(g_{12} + 2g_{23}) \\
& + y_3(g_{32} + g_{20} + g_9 - g_1 - g_{13} - g_{29} + g_4 - g_{24}))e_{34} \\
& + (-2y_0(g_{16} + g_{26}) + y_1(g_1 - g_9 - g_{13} - g_{32} + g_{24} - g_{29} + g_{20} + g_4) \\
& + 2y_2(g_{25} + g_5) + 2y_3(g_6 - g_{22}))e_{23}.
\end{aligned}$$

To get the intersection point of these two lines we take an arbitrary point $X = x\mathbb{R} \in \mathbb{P}^3(\mathbb{R})$ with representation $\mathfrak{X} = x_0e_1 + x_1e_2 + x_2e_3 + x_3e_4 \in \bigwedge^1 \mathcal{G}(\mathbb{P}^3)$ and compute the incidence condition with both lines $\mathfrak{L}_1$ and $\mathfrak{L}_2$:

$$\mathfrak{L}_1 \wedge \mathfrak{X} = 0, \quad \mathfrak{L}_2 \wedge \mathfrak{X} = 0.$$

This results in a system of linear equations for x_0, x_1, x_2, x_3 with solution

$$\begin{aligned}
x_0 =\ & y_0(g_4 + g_9 - g_{29} + g_{13} - g_{20} - g_{32} - g_{24} + g_1) + 2y_1(g_{17} + g_7) \\
& + 2y_2(g_{18} - g_3) + 2y_3(g_2 + g_{19}), \\
x_1 =\ & -2y_0(g_{16} + g_{26}) + y_1(g_1 - g_9 - g_{13} - g_{29} + g_{20} + g_4 - g_{32} + g_{24}) \\
& + 2y_2(g_5 + g_{25}) + 2y_3(g_6 - g_{22}), \\
x_2 =\ & 2y_0(g_{15} - g_{30}) + 2y_1(g_8 + g_{28}) \\
& + y_2(g_{29} - g_{24} - g_4 + g_1 + g_9 + g_{20} - g_{32} - g_{13}) + 2y_3(g_{10} + g_{21}),
\end{aligned}$$

1.6 A Clifford algebraic Approach to Line Geometry

$$x_3 = -2y_0(g_{31}+g_{14})+2y_1(g_{11}-g_{27})$$
$$+ 2y_2(g_{12}+g_{23})+y_3(g_1-g_{32}-g_{20}-g_9+g_{13}+g_{29}-g_4+g_{24}).$$

If we rewrite X as product of a matrix M with the vector $(y_0, y_1, y_2, y_3)^{\mathrm{T}}$ we get

$$(x_0, x_1, x_2, x_3)^{\mathrm{T}} = \mathrm{M} \cdot (y_0, y_1, y_2, y_3)^{\mathrm{T}}, \qquad (1.45)$$

with

$$\mathrm{M} = \begin{pmatrix} k_1 & 2(g_7+g_{17}) & 2(g_{18}-g_3) & 2(g_{19}+g_2) \\ -2(g_{26}+g_{16}) & k_2 & 2(g_5+g_{25}) & 2(g_6-g_{22}) \\ 2(g_{15}-g_{30}) & 2(g_8+g_{28}) & k_3 & 2(g_{21}+2g_{10}) \\ -2(g_{31}+g_{14}) & 2(g_{11}-g_{27}) & 2(g_{23}+g_{12}) & k_4 \end{pmatrix},$$

where

$$k_1 = g_1 - g_{20} - g_{24} - g_{32} - g_{29} + g_9 + g_4 + g_{13},$$
$$k_2 = g_{24} - g_9 + g_{20} - g_{13} - g_{32} + g_1 + g_4 - g_{29},$$
$$k_3 = g_1 - g_{13} - g_{32} - g_4 + g_{29} + g_9 - g_{24} + g_{20},$$
$$k_4 = g_{24} + g_{13} + g_{29} + g_1 - g_4 - g_9 - g_{20} - g_{32}.$$

Therefore, we have found the correspondence between regular collineations and elements of $\mathcal{C}\ell^+_{(3,3,0)}$. For a given versor $\mathfrak{g} \in \mathcal{C}\ell^+_{(3,3,0)}$ corresponding to a collineation we can determine the representation as 4×4 matrix. If we start with a 4×4 matrix A with coefficients (a_{ij}), $i, j = 0, \ldots, 3$ representing a collineation we have to solve a system of 16 linear and 36 quadratic equations to get the corresponding algebra representation. The linear equations are derived with the help of the matrix representation determined above and result in

$$a_{01} = 2(g_7+g_{17}), \quad a_{02} = 2(g_{18}-g_3), \quad a_{03} = 2(g_{19}+g_2), \quad (1.46)$$
$$a_{10} = -2(g_{26}+g_{16}), \quad a_{12} = 2(g_5+g_{25}), \quad a_{13} = 2(g_6-g_{22}),$$
$$a_{20} = 2(g_{15}-g_{30}), \quad a_{21} = 2(g_8+g_{28}), \quad a_{23} = 2(g_{21}+2g_{10}),$$
$$a_{30} = -2(g_{31}+g_{14}), \quad a_{31} = 2(g_{11}-g_{27}), \quad a_{32} = 2(g_{23}+g_{12}),$$

$$a_{00} = g_1 - g_{20} - g_{24} - g_{32} - g_{29} + g_9 + g_4 + g_{13},$$
$$a_{11} = g_{24} - g_9 + g_{20} - g_{13} - g_{32} + g_1 + g_4 - g_{29},$$

$$a_{22} = g_1 - g_{13} - g_{32} - g_4 + g_{29} + g_9 - g_{24} + g_{20},$$
$$a_{33} = g_{24} + g_{13} + g_{29} + g_1 - g_4 - g_9 - g_{20} - g_{32}.$$

These 16 equations have to be solved with the constraint equations derived Eq. (1.43). For both cases $\mathfrak{gg}^* = 1$ and $\mathfrak{gg}^* = -1$ the resulting system of 16 linear and 36 quadratic equations possesses an unique solution. It can be solved analytically with the help of a computer algebra system. We give an example:

Example 1.9. *Let* $K \in \mathrm{PGL}(\mathbb{P}^3(\mathbb{R}))$ *be given by*

$$K = \begin{pmatrix} 1 & 0 & 3 & 0 \\ 1 & 1 & 0 & 1 \\ 1 & 2 & 1 & 0 \\ 1 & 1 & 2 & 1 \end{pmatrix}.$$

To get a versor $\mathfrak{g} \in \mathcal{C}\ell_{(3,3,0)}^+$ *corresponding to this collineation we have to solve the system* (1.46).

$$\begin{aligned}
2(g_7 + g_{17}) &= 0, & 2(g_{18} - g_3) &= 3, & 2(g_{19} + g_2) &= 0, & (1.47)\\
-2(g_{26} + g_{16}) &= 1, & 2(g_5 + g_{25}) &= 0, & 2(g_6 - g_{22}) &= 1,\\
2(g_{15} - g_{30}) &= 1, & 2(g_8 + g_{28}) &= 2, & 2(g_{21} + 2g_{10}) &= 0,\\
-2(g_{31} + g_{14}) &= 1, & 2(g_{11} - g_{27}) &= 1, & 2(g_{23} + g_{12}) &= 2,
\end{aligned}$$

$$\begin{aligned}
g_1 - g_{20} - g_{24} - g_{32} - g_{29} + g_9 + g_4 + g_{13} &= 1,\\
g_{24} - g_9 + g_{20} - g_{13} - g_{32} + g_1 + g_4 - g_{29} &= 1,\\
g_1 - g_{13} - g_{32} - g_4 + g_{29} + g_9 - g_{24} + g_{20} &= 1,\\
g_{24} + g_{13} + g_{29} + g_1 - g_4 - g_9 - g_{20} - g_{32} &= 1.
\end{aligned}$$

We have two possibilities to guarantee that the resulting versor is in the Spin group, i.e., $\mathfrak{gg}^* = 1$ or $\mathfrak{gg}^* = -1$. We compute both solutions and start with the constraint equations implied by Eq. (1.43) and $\mathfrak{gg}^* = 1$. The corresponding Spin group element has the form:

$$\mathfrak{g}_+ = \frac{1}{8\sqrt{2}}(7e_0 + 6e_{12} - 6e_{13} + e_{14} - 2e_{15} - 6e_{23} + 6e_{24} - e_{25} - 2e_{26} + 2e_{34}$$

1.6 A Clifford algebraic Approach to Line Geometry

$$+6e_{35} - 5e_{36} - 4e_{45} + 2e_{46} + 6e_{1234} - 4e_{56} + 6e_{1235} - 6e_{1236}$$
$$-5e_{1245} + 2e_{1246} - 4e_{1256} + 2e_{1345} - e_{1346} + 2e_{1356} - 2e_{2345}$$
$$+2e_{2346} + e_{2356} - 2e_{2456} - e_{123456}).$$

If we demand that $\mathfrak{g}\mathfrak{g}^* = -1$ the resulting Spin group element is computed as

$$\mathfrak{g}_- = \frac{1}{8\sqrt{2}}(e_0 - 6e_{12} - 6e_{13} - e_{14} + 2e_{15} + 4e_{16} + 6e_{23} + 2e_{24} + e_{25} + 2e_{26} + 2e_{34}$$
$$+2e_{35} + 5e_{36} + 2e_{46} - 6e_{1234} + 6e_{1235} + 6e_{1236} + 5e_{1245} - 2e_{1246}$$
$$+6e_{1345} + e_{1346} - 2e_{1356} - 4e_{1456} - 2e_{2345} + 6e_{2346} - e_{2356}$$
$$-2e_{2456} - 4e_{3456} - 7e_{123456}).$$

Both elements $\mathfrak{g}_+$ and $\mathfrak{g}_-$ correspond to the same collineation. With Eq. (1.45) we get the matrix K back with the coefficients of $\mathfrak{g}_+$ and $\mathfrak{g}_-$.

1.6.5 Correlations as Pin Group

We know that grade-1 elements respectively their action on null vectors correspond to null polarities. Thus, correlations are elements of $\mathcal{Cl}^-_{(3,3,0)}$. To transfer a 4×4 matrix representing a correlation into the Clifford algebra we use a similar approach as we used for collineations. An arbitrary field of lines contained in the plane determined by its plane coordinates $X = \mathbb{R}(x_0, x_1, x_2, x_3)^T$ corresponds to the outer product null space of a null three-blade. We apply an arbitrary versor $\mathfrak{h}$ contained in $\mathcal{Cl}^-_{(3,3,0)}$ to this null three-blade. The image is again a null three-blade whose outer product null space is a bundle of lines. Thus, we compute the point concurrent to all lines of this bundle of lines and describe the action of $\mathfrak{h}$ as product of a matrix with a vector. A general element $\mathfrak{h} \in \mathcal{Cl}^-_{(3,3,0)}$ is given by

$$\mathfrak{h} = h_1 e_1 + h_2 e_2 + h_3 e_3 + h_4 e_4 + h_5 e_5 + h_6 e_6 + h_7 e_{123} + h_8 e_{124} + h_9 e_{125} + h_{10} e_{126}$$
$$+ h_{11} e_{134} + h_{12} e_{135} + h_{13} e_{136} + h_{14} e_{145} + h_{15} e_{146} + h_{16} e_{156} + h_{17} e_{234} + h_{18} e_{235}$$
$$+ h_{19} e_{236} + h_{20} e_{245} + h_{21} e_{246} + h_{22} e_{256} + h_{23} e_{345} + h_{24} e_{346} + h_{25} e_{356} + h_{26} e_{456}$$
$$+ h_{27} e_{12345} + h_{28} e_{12346} + h_{29} e_{12356} + h_{30} e_{12456} + h_{31} e_{13456} + h_{32} e_{23456}.$$

If this element shall be a versor we have constraint equations that can be derived from

$$\alpha(\mathfrak{h})v\mathfrak{h}^* \in \bigwedge\nolimits^1 V \text{ for all } v \in \bigwedge\nolimits^1 V. \qquad (1.48)$$

We start with an arbitrary plane given by its plane coordinates $Y = \mathbb{R}(y_0, y_1, y_2, y_3)^T$. Plücker coordinates of three lines contained in this plane can be determined by

$$l_1 = (0:y_3:-y_2:y_0:0:0), \quad l_2 = (-y_3:0,y_1:0:y_0:0), \quad l_3 = (y_2:-y_1:0:0:0:y_0).$$

The null three-blade corresponding to the plane $[l_1, l_2, l_3]$ whose outer product null space is the field of lines defined by l_1, l_2 and l_3 is computed as

$$\begin{aligned}
\mathfrak{b} &= (y_3 e_2 - y_2 e_3 + y_0 e_4) \wedge (-y_3 e_1 + y_1 e_3 + y_0 e_5) \wedge (y_2 e_1 - y_1 e_2 + y_0 e_6) \\
&= -y_0 y_1 y_2 e_{134} + y_0 y_3 y_2 e_{125} - y_2^2 y_0 e_{135} + y_0^2 y_2 e_{145} + y_0 y_3 y_1 e_{124} \\
&\quad + y_0 y_1^2 e_{234} + y_0 y_1 y_2 e_{235} - y_0^2 y_1 e_{245} + y_3^2 y_0 e_{126} - y_0 y_3 y_2 e_{136} \\
&\quad + y_0^2 y_3 e_{146} + y_0 y_3 y_1 e_{236} - y_0^2 y_1 e_{346} + y_0^2 y_3 e_{256} - y_0^2 y_2 e_{356} + y_0^3 e_{456}.
\end{aligned}$$

We apply the sandwich operator with the general element $\mathfrak{h} \in \mathcal{C}\ell^-_{(3,3,0)}$. Since a versor from the odd part of the algebra corresponds to a correlation the null three-blade $\mathfrak{b}' = \alpha(\mathfrak{h})\mathfrak{b}\mathfrak{h}^*$ corresponds to a bundle of lines. Thus, we can apply the same procedure that we used for collineations to determine the image point, *i.e.*, the point concurrent to all lines of the bundle of lines. Therefore, we compute the outer product null space of $\mathfrak{b}'$

$$\mathbb{NO}(\mathfrak{b}') = \{\alpha v_1 + \beta v_2 + \gamma v_3 | \alpha, \beta, \gamma \in \mathbb{R}\},$$

where

$$\begin{aligned}
v_1 = &- \bigl(2 y_0 h_7 + y_1(h_{29} - h_1 - h_9 - h_{13}) + y_2(h_8 - h_2 - h_{19} - h_{28}) \\
&+ y_3(h_{11} + h_{27} + h_{18} - h_3)\bigr) e_1 \\
&- \bigl(y_0(h_3 + h_{11} - h_{27} + h_{18}) + y_1(h_{25} + h_{14} + h_{31} - h_5) \\
&+ y_2(h_4 + h_{32} - h_{24} + h_{20}) + 2 y_3 h_{23}\bigr) e_5 \\
&+ \bigl(y_0(h_2 + h_8 - h_{19} + h_{28}) + y_1(h_6 + h_{22} - h_{15} + h_{30}) - 2 y_2 h_{21} \\
&+ y_3(h_{20} - h_4 - h_{32} - h_{24})\bigr) e_6,
\end{aligned}$$

1.6 A Clifford algebraic Approach to Line Geometry   69

$$\mathfrak{v}_2 = -\bigl(2y_0h_7 + y_1(h_{29} - h_1 - h_9 - h_{13}) + y_2(h_8 - h_2 - h_{28} - h_{19})$$
$$+ y_3(h_{27} + h_{18} - h_3 + h_{11})\bigr)e_2$$
$$+ \bigl(y_0(h_3 + h_{11} + h_{18} - h_{27}) + y_1(h_{25} + h_{14} + h_{31} - h_5)$$
$$+ y_2(h_4 + h_{32} - h_{24} + h_{20}) + 2y_3h_{23}\bigr)e_4$$
$$- \bigl(y_0(h_1 - h_{29} - h_{13} - h_9) + 2h_{16}y_1 + y_2(h_{22} - h_6 - h_{15} - h_{30})$$
$$+ y_3(h_{25} + h_5 - h_{31} + h_{14})\bigr)e_6,$$

$$\mathfrak{v}_3 = -\bigl(2y_0h_7 + y_1(h_{29} - h_{13} - h_1 - h_9) + y_2(h_8 - h_2 - h_{28} - h_{19})$$
$$+ y_3(h_{11} - h_3 + h_{27} + h_{18})\bigr)e_3$$
$$- \bigl(y_0(h_2 + h_8 - h_{19} + h_{28}) + y_1(h_6 + h_{22} - h_{15} + h_{30}) - 2y_2h_{21}$$
$$+ y_3(h_{20} - h_4 - h_{32} - h_{24})\bigr)e_4$$
$$+ \bigl(y_0(h_1 - h_{29} - h_{13} - h_9) + 2h_{16}y_1 + y_2(h_{22} - h_6 - h_{15} - h_{30})$$
$$+ y_3(h_5 + h_{14} + h_{25} - h_{31})\bigr)e_5.$$

These three null vectors correspond to three points on Klein's quadric that span the bundle of lines defined by the outer product null space of $\mathfrak{b}'$. The point concurrent to these three lines is computed with the help of the exterior algebra $\mathcal{G}(\mathbb{P}^3)$. Hence, we transfer the null vectors $\mathfrak{v}_1, \mathfrak{v}_2$ to two-blades $\mathfrak{L}_1, \mathfrak{L}_2$ of $\mathcal{G}(\mathbb{P}^3)$ with Eq. (1.44). The intersection point of these two lines is the image of the plane $Y = \mathbb{R}(y_0, y_1, y_2, y_3)^\mathrm{T}$. A general point $X = (x_0, x_1, x_2, x_3)^\mathrm{T}\mathbb{R}$ is written as element of the exterior algebra $\mathfrak{X} = x_0e_1 + x_1e_2 + x_2e_3 + x_3e_4 \in \mathcal{G}(\mathbb{P}^3)$. The point X is incident with the lines L_1 and L_2 if $\mathfrak{L}_1 \wedge \mathfrak{X} = 0$ and $\mathfrak{L}_2 \wedge \mathfrak{X} = 0$. This results in a system of linear equations for x_0, x_1, x_2, x_3 with solution

$$x_0 = -2y_0h_7 + y_1(h_1 + h_9 + h_{13} - h_{29}) + y_2(h_2 + h_{19} + h_{28} - h_8)$$
$$+ y_3(h_3 - h_{18} - h_{27} - h_{11}),$$
$$x_1 = y_0(h_{13} - h_1 + h_{29} + h_9) - 2y_1h_{16} + y_2(h_6 + h_{30} + h_{15} - h_{22})$$
$$+ y_3(h_{31} - h_{25} - h_{14} - h_5),$$
$$x_2 = y_0(h_{19} - h_2 - h_8 - h_{28}) + y_1(h_{15} - h_{30} - h_6 - h_{22}) + 2h_{21}y_2$$
$$+ y_3(h_4 - h_{20} + h_{32} + h_{24}),$$
$$x_3 = y_0(h_{27} - h_3 - h_{11} - h_{18}) + y_1(h_5 - h_{31} - h_{25} - h_{14})$$
$$+ y_2(h_{24} - h_4 - h_{20} - h_{32}) - 2h_{23}y_3.$$

If we rewrite X as a product of a matrix C with $Y = (y_0, y_1, y_2, y_3)^T$ we get:

$$\begin{pmatrix} x_0 \\ x_1 \\ x_2 \\ x_3 \end{pmatrix} = \begin{pmatrix} c_{00} & c_{01} & c_{02} & c_{03} \\ c_{10} & c_{11} & c_{12} & c_{13} \\ c_{20} & c_{21} & c_{22} & c_{23} \\ c_{30} & c_{31} & c_{32} & c_{33} \end{pmatrix} \begin{pmatrix} y_0 \\ y_1 \\ y_2 \\ y_3 \end{pmatrix}, \text{ with} \quad (1.49)$$

$c_{01} = h_1 + h_9 + h_{13} - h_{29}, \quad c_{02} = h_2 + h_{19} + h_{28} - h_8, \quad c_{03} = h_3 - h_{18} - h_{27} - h_{11},$
$c_{10} = h_{13} - h_1 + h_{29} + h_9, \quad c_{12} = h_6 + h_{30} + h_{15} - h_{22}, \quad c_{13} = h_{31} - h_{25} - h_{14} - h_5,$
$c_{20} = h_{19} - h_2 - h_8 - h_{28}, \quad c_{21} = h_{15} - h_{30} - h_6 - h_{22}, \quad c_{23} = h_4 - h_{20} + h_{32} + h_{24},$
$c_{30} = h_{27} - h_3 - h_{11} - h_{18}, \quad c_{31} = h_5 - h_{31} - h_{25} - h_{14}, \quad c_{32} = h_{24} - h_4 - h_{20} - h_{32},$

and

$$c_{00} = -2h_7, \quad c_{11} = -2h_{16}, \quad c_{22} = 2h_{21}, \quad c_{33} = -2h_{23}.$$

The matrix C describes the correspondence between regular correlations and elements of $\mathcal{C}\ell^-_{(3,3,0)}$. For a given versor that corresponds to a correlation, we can compute the matrix representation by Eq. (1.49). Furthermore, we can determine the versor corresponding to a regular correlation represented by a 4×4 matrix A if we solve the system of 16 linear equations given by

$a_{01} = h_1 + h_9 + h_{13} - h_{29}, \quad a_{02} = h_2 + h_{19} + h_{28} - h_8, \quad a_{03} = h_3 - h_{18} - h_{27} - h_{11},$
$a_{10} = h_{13} - h_1 + h_{29} + h_9, \quad a_{12} = h_6 + h_{30} + h_{15} - h_{22}, \quad a_{13} = h_{31} - h_{25} - h_{14} - h_5,$
$a_{20} = h_{19} - h_2 - h_8 - h_{28}, \quad a_{21} = h_{15} - h_{30} - h_6 - h_{22}, \quad a_{23} = h_4 - h_{20} + h_{32} + h_{24},$
$a_{30} = h_{27} - h_3 - h_{11} - h_{18}, \quad a_{31} = h_5 - h_{31} - h_{25} - h_{14}, \quad a_{32} = h_{24} - h_4 - h_{20} - h_{32},$

$$a_{00} = -2h_7, \quad a_{11} = -2h_{16}, \quad a_{22} = 2h_{21}, \quad a_{33} = -2h_{23} \quad (1.50)$$

with respect to the constraint equations derived by the condition $\mathfrak{h}\mathfrak{h}^* = 1$ or $\mathfrak{h}\mathfrak{h}^* = -1$, see Eq. (1.48). With the help of a computer algebra system it can be verified that this system possesses an unique solution for each of the cases $\mathfrak{h}\mathfrak{h}^* = \pm 1$.

Example 1.10. *As example we take again the matrix*

$$K = \begin{pmatrix} 1 & 0 & 3 & 0 \\ 1 & 1 & 0 & 1 \\ 1 & 2 & 1 & 0 \\ 1 & 1 & 2 & 1 \end{pmatrix},$$

1.6 A Clifford algebraic Approach to Line Geometry

but now the matrix describes a correlation. To get the corresponding versor we have to solve the system of linear equations

$$h_1+h_9+h_{13}-h_{29}=0, \quad h_2+h_{19}+h_{28}-h_8=3, \quad h_3-h_{18}-h_{27}-h_{11}=0,$$
$$h_{13}-h_1+h_{29}+h_9=1, \quad h_6+h_{30}+h_{15}-h_{22}=0, \quad h_{31}-h_{25}-h_{14}-h_5=1,$$
$$h_{19}-h_2-h_8-h_{28}=1, \quad h_{15}-h_{30}-h_6-h_{22}=2, \quad h_4-h_{20}+h_{32}+h_{24}=0,$$
$$h_{27}-h_3-h_{11}-h_{18}=1, \quad h_5-h_{31}-h_{25}-h_{14}=1, \quad h_{24}-h_4-h_{20}-h_{32}=2,$$

$$-2h_7=1, \quad -2h_{16}=1, \quad 2h_{21}=1, \quad -2h_{23}=1.$$

If we solve this system of linear equations with respect to the constraints given by Eq. (1.48) and $\mathfrak{h}\mathfrak{h}^*=1$ *the resulting versor is given by*

$$\mathfrak{h}_+ = \frac{1}{8}(e_1+e_2+e_3+2e_4-4e_6-2e_{123}-3e_{124}+e_{125}+4e_{126}-e_{134}+e_{136}$$
$$-2e_{145}-2e_{146}-2e_{156}-2e_{234}-e_{235}+5e_{236}+2e_{246}-4e_{245}-2e_{345}$$
$$-6e_{256}+4e_{456}-2e_{356}+3e_{12345}+3e_{12346}+3e_{12356}-6e_{23456}).$$

For $\mathfrak{h}\mathfrak{h}^*=-1$ *we get*

$$\mathfrak{h}_- = \frac{1}{8}(-3e_1+3e_2-3e_3-6e_4-2e_{123}-5e_{124}+e_{125}-4e_{126}-e_{134}+e_{136}$$
$$-2e_{145}+6e_{146}-2e_{156}+2e_{234}-e_{235}+3e_{236}+2e_{246}-2e_{345}+2e_{256}$$
$$+4e_{346}-4e_{456}-2e_{356}-e_{12345}+e_{12346}-e_{12356}-4e_{12456}+2e_{23456}).$$

Note that both versors $\mathfrak{h}_+$ *and* $\mathfrak{h}_-$ *correspond to the same correlation. The correlation can be computed with Eq. (1.49).*

Remark 1.16. With the knowledge which elements correspond to collineations and which correspond to correlations we can apply a transformation to a pencil of lines to study the action of a mapping on a point. Since a general projective transformation maps pencils of lines to pencils of lines, we could not distinguish between collineations and correlations. Therefore, we used bundles of lines and fields of lines to study the action of a general projective transformation represented as element of $C\ell_{(3,3,0)}$.

1.6.6 Singular projective Transformations

Now we examine singular projective transformations. Every versor can be written as the geometric product of grade-1 elements corresponding to null polarities. The square of a grade-1 element (see Eq. (1.37)) is related to the determinant of the corresponding null polarity, see Eq. (1.41). Thus, we assume that one of these null polarities is a singular one, and therefore, its determinant is equal to zero. We define:

Definition 1.18. *An element* $\mathfrak{g} \in \mathcal{Cl}_{(3,3)}$ *with* $\mathfrak{g} = \mathfrak{v}_1 \ldots \mathfrak{v}_k$ *with* $k \leq 6$ *and* $\mathfrak{v}_i \in \bigwedge^1 V$ *is called a* null versor *if at least one* $\mathfrak{v}_i$ *squares to zero.*

Constraint equations can be derived from

$$\alpha(\mathfrak{g})\mathfrak{v}\mathfrak{g}^* \in \bigwedge^1 V \text{ for all } \mathfrak{v} \in \bigwedge^1 V, \quad \mathfrak{g} \in \mathcal{Cl}^+_{(3,3)}$$

for a singular collineation and

$$\alpha(\mathfrak{h})\mathfrak{v}\mathfrak{h}^* \in \bigwedge^1 V \text{ for all } \mathfrak{v} \in \bigwedge^1 V, \quad \mathfrak{h} \in \mathcal{Cl}^-_{(3,3)}$$

for a singular correlation. The matrix representations for general collineations (1.45) and correlations (1.49) are also valid for the singular case. The system of linear equations (1.46) with the constraint equations implied by $\alpha(\mathfrak{g})\mathfrak{v}\mathfrak{g}^* \in \bigwedge^1 V$ for all $\mathfrak{v} \in \bigwedge^1 V$ with $\mathfrak{g} \in \mathcal{Cl}^+_{(3,3)}$ can not be solved for a singular collineation. For a singular correlation the system of linear equations (1.50) with the constraint equations implied by $\alpha(\mathfrak{h})\mathfrak{v}\mathfrak{h}^* \in \bigwedge^1 V$ for all $\mathfrak{v} \in \bigwedge^1 V$ with $\mathfrak{h} \in \mathcal{Cl}^-_{(3,3)}$ has also no solution. Nevertheless, we are able to write singular correlations or collineations as null versors if we know the null polarities that generate them. These null polarities can be transferred to vectors in the homogeneous Clifford algebra model, and therefore, the whole correlation or collineation can be transferred to the algebra model.

1.7 A Clifford algebraic Approach to Lie Sphere Geometry

The same construction that we described for Klein's quadric can be applied to any quadric. As a second example that shall demonstrate

1.7 A Clifford algebraic Approach to Lie Sphere Geometry

the power of this calculus we examine Lie sphere geometry in a Clifford algebra context. Lie sphere geometry is the geometry of oriented spheres. Especially, for the three-dimensional case the set of oriented spheres can be mapped to a hyperquadric L_1^4 in five-dimensional projective space $\mathbb{P}^5(\mathbb{R})$. The construction goes back to S. LIE and was treated again by W. BLASCHKE, cf. [8]. A modern treatment of this topic can be found in [14]. Moreover, the Lie construction can be achieved for arbitrary dimension.

1.7.1 Lie's Quadric

A point model for the set of oriented hyperspheres, hyperplanes, and points (considered as spheres of radius 0) of $\mathbb{R}^n$ is given by the projective quadric

$$L_1^{n+1} : -x_0^2 + x_1^2 + \ldots + x_{n+1}^2 - x_{n+2}^2 = 0.$$

For our purposes it is convenient that we restrict ourselves to the case of oriented spheres in three-dimensional Euclidean space. Nevertheless, we formulate the calculus for arbitrary dimensions. The quadric $L_1^{n+1} \subset \mathbb{P}^{n+2}(\mathbb{R})$ is of dimension $n+1$, degree 2, and is called Lie's quadric. The maximal dimension of subspaces contained by L_1^{n+1} is 1, and therefore, there are no two-spaces contained entirely in L_1^{n+1}. Oriented hyperspheres, hyperplanes, and points are represented in Lie coordinates as shown in Table 1.3. It is not difficult to recover the Euclidean representation from Lie coordinates. If $x_0 + x_1 = 0$ and

Table 1.3: Correspondence between Euclidean entities and Lie-coordinates.

Euclidean	Lie
points: $u \in \mathbb{R}^n$	$\left(\frac{1+u \cdot u}{2}, \frac{1-u \cdot u}{2}, u_1, \ldots, u_n, 0\right)^T \mathbb{R}$
∞	$(1, -1, 0, \ldots, 0, 0)^T \mathbb{R}$
sphere: center $p \in \mathbb{R}^n$, signed radius r	$\left(\frac{1+p \cdot p - r^2}{2}, \frac{1-p \cdot p + r^2}{2}, p_1, \ldots, p_n, r\right)^T \mathbb{R}$
planes: $u \cdot N = h$, unit normal $N \in \mathbb{R}^n$	$(h, -h, N_1, \ldots, N_n, 1)^T \mathbb{R}$

if $x_{n+2} = 0$ we have the point at infinity. If $x_{n+2} \neq 0$ we bring the point to the form $(h, -h, N, 1)^T \mathbb{R}$ by dividing by x_{n+2}. If $x_0 + x_1 \neq 0$ and if $x_{n+2} = 0$, we have a proper point. We obtain its normal form by dividing by $x_0 + x_1$. The last case is if $x_{n+2} \neq 0$. In this case we have an oriented sphere. Again we get its normal form through division by $x_0 + x_1$.

The fundamental invariant of Lie sphere geometry is the oriented contact of spheres. It is not difficult to show that two spheres are in oriented contact if, and only if, their Lie coordinates $s_1, s_2 \in L_1^{n+1}$ satisfy $\ell(s_1, s_2) = 0$, where $\ell(\cdot, \cdot)$ denotes the bilinear form corresponding to L_1^{n+1}.

Especially for $n = 3$ the lines on L_1^4 correspond to so called parabolic pencils of spheres. These pencils consist of all oriented spheres with one common point of contact. Furthermore, each parabolic pencil contains exactly one point, *i.e.*, sphere of radius 0. If this point sphere is not ∞ the pencil contains exactly one oriented hyperplane Σ.

Remark 1.17. *Conics on Lie's quadric correspond to Dupin cyclides, that are the envelopes of two one-parameter families of spheres.*

The group of Lie transformations shows up as the group of projective automorphisms of L_1^{n+1}. This group is isomorphic to $O(n+1, 2)/\pm 1$, see CECIL [14]. Since the Pin group of the Clifford algebra $\mathcal{C}\ell_{(n+1,2,0)}$ is a double cover of $O(n+1, 2)$ we can use this group to describe Lie transformations.

1.7.2 The homogeneous Clifford Algebra Model corresponding to Lie Sphere Geometry

In this section we discuss the Clifford algebra model for Lie Sphere Geometry in the three-dimensional case. Therefore, the projective space we are dealing with is a five-dimensional space $\mathbb{P}^5(\mathbb{R})$. The homogeneous Clifford algebra model is obtained with the six-dimensional

1.7 A Clifford algebraic Approach to Lie Sphere Geometry

real vector space $\mathbb{R}^6$ as a model for the projective image space together with the quadratic form of Lie's quadric

$$Q = \begin{pmatrix} -1 & 0 & 0 & 0 & 0 & 0 \\ 0 & 1 & 0 & 0 & 0 & 0 \\ 0 & 0 & 1 & 0 & 0 & 0 \\ 0 & 0 & 0 & 1 & 0 & 0 \\ 0 & 0 & 0 & 0 & 1 & 0 \\ 0 & 0 & 0 & 0 & 0 & -1 \end{pmatrix}.$$

This algebra has signature $(p,q,r) = (4,2,0)$ and is of dimension $2^6 = 64$. Again the advantage of the Clifford algebra lies in the common description of the application of Lie transformations. Arbitrary projective subspaces of $\mathbb{P}^5(\mathbb{R})$ are transformed by the sandwich operator. As an example we determine all Lie inversions that leave the point at infinity fixed, i.e., the subgroup of *Laguerre* transformations. The point at infinity has the form $\mathfrak{p} = e_1 - e_2$, compare to Table 1.3. A general invertible grade-1 element is given by

$$\mathfrak{a} = a_1 e_1 + a_2 e_2 + a_3 e_3 + a_4 e_4 + a_5 e_5 + a_6 e_6, \quad -a_1^2 + a_2^2 + a_3^2 + a_4^2 + a_5^2 - a_6^2 \neq 0.$$

The application of the sandwich operator to $\mathfrak{p}$ results in

$$\begin{aligned}\alpha(\mathfrak{a})\mathfrak{p}\mathfrak{a}^{-1} = &-2(a_3^2 + a_4^2 + a_1^2 + a_2^2 - a_6^2 + 2a_2 a_1 + a_5^2) e_1 \\ &- 2(-a_3^2 - a_4^2 + a_6^2 + a_1^2 + a_2^2 + 2a_2 a_1 - a_5^2) e_2 - 4(a_1 + a_2) a_3 e_3 \\ &- 4(a_1 + a_2) a_4 e_4 - 4(a_1 + a_2) a_5 e_5 - 4(a_1 + a_2) a_6 e_6.\end{aligned}$$

To guarantee that this entity represents the point at infinity, we first see that $a_1 + a_2 = 0$. With this condition the coefficients of e_3, e_4, e_5, and e_6 vanish. Moreover, the sum of the coefficients e_1 and e_2 has to vanish. This results in

$$-4a_1^2 - 4a_2^2 - 8a_2 a_1 = -4(a_1 + a_2)^2 = 0.$$

Therefore, the only condition to a Lie inversion that it represents a Laguerre transformation is given by $a_1 + a_2 = 0$ and the subgroup of Laguerre transformations is generated by all vectors with $a_1 + a_2 = 0$.

Remark 1.18. *Analogue to Th. 1.3 we can formulate a similar theorem for Lie sphere geometry in arbitrary dimensions. Since the* projective model space for n-dimensional Lie sphere geometry has dimension $n+2$, the vector space for the homogeneous Clifford algebra model has dimension $n+3$. That means the highest grade is equal to $n+3$, and therefore, every group element can be written as the composition of $n+3$ grade-1 elements at the most. Especially for the case $n=3$, we have similar results as for the Klein quadric. In this case six involutions are necessary to generate the whole group.

Let us reformulate this remark as theorem.

Theorem 1.4. *Every Lie transformation in n-dimensional space is the composition of $n+3$ involutions, that correspond to the sandwich action of grade-1 elements.*

1.8 A Clifford algebraic Approach to Study's Quadric

Study's quadric was discussed in section 1.2.2. To obtain a Clifford algebra description of the Study model we take the eight-dimensional real vector space $\mathbb{R}^8$ as a model for the seven-dimensional real projective space $\mathbb{P}^7(\mathbb{R})$ together with the quadratic form corresponding to Study's quadric

$$Q = \begin{pmatrix} O & I \\ I & O \end{pmatrix},$$

where I denotes the 4×4 identity matrix and O the 4×4 zero matrix. The corresponding Clifford algebra has signature $(p,q,r) = (4,4,0)$ and is of dimension $2^8 = 256$. Null vectors correspond to points on Study's quadric. Subspaces of $\mathbb{P}^7(\mathbb{R})$ of higher dimensions can be obtained by the wedge product of grade-1 elements, as in the Grassmann algebra $\bigwedge \mathbb{P}^7(\mathbb{R})$. The additional inner product structure allows us to characterize points on the quadric as null vectors. With the sandwich operator it is possible to apply automorphisms of Study's quadric to subspaces. Especially, conics defined by intersections of two-spaces with the quadric can be transformed at once.

1.9 A Clifford algebraic Approach to Study's Sphere

The same construction that was done for Klein's, Lie's, and Study's quadric can also be applied to Study's sphere, since it is a quadric over the commutative ring of dual numbers. Furthermore, it is possible to construct an affine and a projective model for Study's sphere $S_{\mathbb{D}}^2$.

Projective Model The Clifford algebra is defined by the module $\mathcal{M} = \mathbb{D}^4$ that serves as model for the projective space $\mathbb{P}^3(\mathbb{D})$ and the quadratic form $Q = M$ belonging to the unit sphere

$$S_{\mathbb{D}}^2 : x^T M x = 0, \text{ with } X = x\mathbb{D}^\times \in \mathbb{P}^3(\mathbb{D}) \text{ and } M = \begin{pmatrix} -1 & 0 & 0 & 0 \\ 0 & 1 & 0 & 0 \\ 0 & 0 & 1 & 0 \\ 0 & 0 & 0 & 1 \end{pmatrix}.$$

Oriented lines are represented by null vectors.

Affine Model The affine model is obtained by using the module $\mathcal{M} = \mathbb{D}^3$ and $Q = \begin{pmatrix} 1 & 0 & 0 \\ 0 & 1 & 0 \\ 0 & 0 & 1 \end{pmatrix}$. In this case oriented lines are represented by grade-1 elements that square to 1 respectively Pin group elements of the $\bigwedge^1 V$ subspace.

Remark 1.19. *To obtain a Clifford algebra model of Study's sphere of oriented displacements we have to increase the dimensions by one. Different dual unit quaternions that describe the same displacement correspond to different points in this model. For the projective model we need* $\mathcal{M} = \mathbb{D}^5$ *as model for* $\mathbb{P}^4(\mathbb{D})$ *and*

$$Q = \begin{pmatrix} -1 & 0 & 0 & 0 & 0 \\ 0 & 1 & 0 & 0 & 0 \\ 0 & 0 & 1 & 0 & 0 \\ 0 & 0 & 0 & 1 & 0 \\ 0 & 0 & 0 & 0 & 1 \end{pmatrix}$$

as the matrix of the quadratic form for the projective model. Oriented displacements correspond to null vectors in this model. For the affine model we need $\mathcal{M} = \mathbb{D}^4$ and

$$Q = \begin{pmatrix} 1 & 0 & 0 & 0 \\ 0 & 1 & 0 & 0 \\ 0 & 0 & 1 & 0 \\ 0 & 0 & 0 & 1 \end{pmatrix}.$$

In this case oriented displacements correspond to grade-1 elements that square to 1 respectively to Pin group elements from the $\bigwedge^1 V$ subspace.

1.10 Quadric Geometric Algebra

In this section we present a new geometric algebra model that was introduced by Zamora-Esquivel [68]. With this model a generalization of the conformal model is achieved. We discuss the geometric objects that can be represented. We extend this model and identify the Pin group of this geometric algebra as the group of inversions with respect to quadrics in principal axis position except for a translation. Moreover, we discuss the construction for the two- and three-dimensional case in detail and give the construction for arbitrary dimension. We start with the planar, *i.e.*, two-dimensional case before we move on to higher dimensions.

1.10.1 The Embedding

The quadric geometric algebra for the two-dimensional case is constructed with a Clifford algebra over a six-dimensional vector space $V = \mathbb{R}^6$. In fact the term quadric could be replaced by the term conic for the two-dimensional case. Without loss of generality we call it quadric geometric algebra and abbreviate this term with QnGA, where n denotes the dimension of the base space. The quadratic form we are using is derived by the quadratic form of the conformal geometric algebra used in [18]:

1.10 Quadric Geometric Algebra

$$Q = \begin{pmatrix} 0 & 0 & -1 & 0 & 0 & 0 \\ 0 & 1 & 0 & 0 & 0 & 0 \\ -1 & 0 & 0 & 0 & 0 & 0 \\ 0 & 0 & 0 & 0 & 0 & -1 \\ 0 & 0 & 0 & 0 & 1 & 0 \\ 0 & 0 & 0 & -1 & 0 & 0 \end{pmatrix}.$$

The signature of the resulting algebra is $(p,q,r) = (4,2,0)$. For every axis, i.e., the x- and the y-axis a conformal embedding is performed, see [19]. Therefore, we have the embedding $\eta : \mathbb{R}^2 \to \bigwedge^1 V$,

$$\mathbb{R}^2 \ni P \mapsto \mathfrak{p} \in \bigwedge^1 V,$$

$$P = (x,y)^\mathrm{T} \mapsto e_1 + xe_2 + \frac{1}{2}x^2 e_3 + e_4 + ye_5 + \frac{1}{2}y^2 e_6 = \mathfrak{p}. \qquad (1.51)$$

Affine points $P = (x,y)^\mathrm{T} \in \mathbb{R}^2$ are embedded as null vectors. This means

$$\eta(P)^2 = 0 \text{ for } P \in \mathbb{R}^2. \qquad (1.52)$$

The projection on the generator subspace spanned by e_1, e_2, and e_3 is denoted by subscript x and the projection on e_4, e_5, e_6 by subscript y. Due to the fact, that the embedding is conformal (see [19, 68])for both axes we get the additional conditions:

$$\eta(P)_x^2 = 0, \quad \eta(P)_y^2 = 0. \qquad (1.53)$$

In the following we call grade-1 elements satisfying (1.52) and (1.53) *embedded points*. Let $P_1 = (x_1, y_1)^\mathrm{T} \in \mathbb{R}^2$ and $P_2 = (x_2, y_2)^\mathrm{T} \in \mathbb{R}^2$ be two point. The inner product of their images under η results in

$$\eta(P_1) \cdot \eta(P_2) = -\frac{1}{2}x_2^2 + x_1 x_2 - \frac{1}{2}x_1^2 - \frac{1}{2}y_2^2 + y_1 y_2 - \frac{1}{2}y_1^2$$

$$= -\frac{1}{2}\left((x_2 - x_1)^2 + (y_2 - y_1)^2\right)$$

$$= -\frac{1}{2}d_E^2(P_1, P_2),$$

where $d_E(P_1, P_2)$ denotes the Euclidean distance between the points P_1 and P_2. Note that this formula only is true for normalized null vectors. This means that the homogeneous factors have to be equal to one. These vectors can be interpreted as images of points under η and we call them normalized. They are characterized by

$$-e_3 \cdot \mathfrak{p} = 1, \quad -e_6 \cdot \mathfrak{p} = 1. \tag{1.54}$$

The inner product of an arbitrary embedded point with e_3 or e_6 is constant. Therefore, we can interpret these elements as points at infinity. Furthermore, the combination of the conditions (1.54) results in

$$-(e_3 + e_6) \cdot \mathfrak{p} = 2.$$

Thus, the geometric entity corresponding $e_3 + e_6$ can be interpreted as point at infinity. The elements e_3 and e_6 represent the ideal points corresponding to each axis and $e_3 + e_6$ represents a point at infinity contained in both axes. Geometrically, these three algebra elements describe the same point, *i.e.*, the point at infinity ∞ although they differ algebraically. There are grade-1 elements that satisfy the conditions (1.52) and (1.53) without having a pre-image in $\mathbb{R}^2$. For example $e_3, e_6, e_3 + e_6$ and algebra elements of the form:

$$\mathfrak{u}_1 = e_1 + x_0 e_2 + \frac{1}{2} x_0^2 e_3 + e_6, \quad \mathfrak{u}_2 = e_3 + e_4 + y_0 e_5 + \frac{1}{2} y_0^2 e_6.$$

If we determine the Euclidean distance of an embedded point to $\mathfrak{u}_1$ or $\mathfrak{u}_2$, the result is a complex number and depends on x_0 respectively y_0. Hence, these elements do not represent points.

1.10.2 Geometric Entities

To calculate the pre-image $\eta^{-1}(\mathfrak{p})$ of $\mathfrak{p} \in \bigwedge^1 V$ representing an embedded point, *i.e.*, an algebra element fulfilling (1.52) and (1.53), we recall the definition of the GIPNS and the GOPNS, see [53].

The *geometric inner product null space* (GIPNS) and dual the *geometric outer product null space* (GOPNS) of a k-blade $\mathfrak{A} \in \bigwedge^k V$ in Q2GA is defined by

$$\mathbb{NI}_G(\mathfrak{A}) := \left\{ (x, y)^T \in \mathbb{R}^2 : \eta(x, y) \cdot \mathfrak{A} = 0 \right\},$$
$$\mathbb{NO}_G(\mathfrak{A}) := \left\{ (x, y)^T \in \mathbb{R}^2 : \eta(x, y) \wedge \mathfrak{A} = 0 \right\}.$$

1.10 Quadric Geometric Algebra

Remark 1.20. *When dealing with an algebra element and the corresponding geometric entity, we mention explicitly what null space is meant. For example we talk about inner product conics. This means the inner product null space defines a conic in $\mathbb{R}^2$.*

Before we start the examination of geometric objects occurring in this model we define special 5-blades that are necessary to change from inner product to outer product null spaces and vice versa.

Definition 1.19. *On the one hand the 5-blade*

$$\mathfrak{J} = e_2 \wedge e_5 \wedge e_1 \wedge e_4 \wedge (e_3 + e_6)$$

maps outer product null spaces to inner product null spaces

$$\mathfrak{J} : \bigwedge^i V \to \bigwedge^{k-i} V,$$
$$\bigwedge^i V \ni \mathfrak{v} \mapsto \mathfrak{v} \cdot \mathfrak{J} \in \bigwedge^{k-i} V,$$

with $i \in \{1, \ldots, 4\}$ and $k = 5$ for the planar quadric geometric algebra. On the other hand

$$\mathfrak{J}^* := e_2 \wedge e_5 \wedge e_3 \wedge e_6 \wedge (e_1 + e_4)$$

maps dual elements to normal elements respectively inner product null spaces to outer product null spaces. There is no difference in left- or right multiplication with these 5-blades. The result differs by the factor -1 and describes the same geometric entity.

With Def. 1.19 we get (see [68])

$$\mathbb{NI}_G(\mathfrak{A}) = \mathbb{NO}_G(\mathfrak{A} \cdot \mathfrak{J}),$$
$$\mathbb{NO}_G(\mathfrak{A}) = \mathbb{NI}_G(\mathfrak{A} \cdot \mathfrak{J}^*).$$

Note that dualization is realized with the inner product. Now we take a look at the inner product null space of grade-1 elements that are not embedded points. Therefore, at least one of the conditions (1.52) or (1.53) is not satisfied. Let $\mathfrak{c} = -2a_1 e_1 + 2a_2 e_2 - a_3 e_3 - 2a_4 e_4 + 2a_5 e_5 - a_6 e_6$ be a general grade-1 element. The GIPNS results

$$\mathbb{NI}_G(\mathfrak{c}) = \{(x,y)^T \in \mathbb{R}^2 \mid \eta(x,y) \cdot \mathfrak{c} = 0\}$$
$$= \{(x,y)^T \in \mathbb{R}^2 \mid a_1 x^2 + 2a_2 x + a_3 + a_4 y^2 + 2a_5 y + a_6 = 0\}.$$

The GIPNS is a conic in principal position, because there is no term containing xy. Any conic has a coefficient matrix and is given by

$$\begin{pmatrix} 1 & x & y \end{pmatrix} \begin{pmatrix} a_{11} & a_{12} & a_{13} \\ a_{12} & a_{22} & a_{23} \\ a_{13} & a_{23} & a_{33} \end{pmatrix} \begin{pmatrix} 1 \\ x \\ y \end{pmatrix} = 0.$$

Therefore, we can define a bijection χ between those symmetric matrices which represent conics in principal position and vectors by

$$\begin{pmatrix} a_0 & a_2 & a_5 \\ a_2 & a_1 & 0 \\ a_5 & 0 & a_4 \end{pmatrix} \mapsto 2a_1 e_1 - 2a_2 e_2 + a_3 e_3 + 2a_4 e_4 - 2a_5 e_5 + a_6 e_6.$$

For the bijection (1.55) we assume that $a_3 := \frac{1}{2} a_0$ and $a_6 := \frac{1}{2} a_0$. It would be sufficient to demand that $a_3 + a_6 = a_0$ to result in the same conic, because the constant value is equal to $a_3 + a_6$. This does not change the GIPNS of the conic. With Eq. (1.55) embedded points can be interpreted as circles whose radii are equal to zeor.

After dualization an inner product conic $\mathfrak{c}$ becomes an outer product conic $\hat{\mathfrak{c}}$ that is a four-blade and can be generated by the outer product of four embedded points. Hence, these four points lie on the conic because

$$\mathfrak{p}_i \in \mathbb{NO}_G(\mathfrak{p}_1 \wedge \mathfrak{p}_2 \wedge \mathfrak{p}_3 \wedge \mathfrak{p}_4), \text{ for } i = 1, \dots, 4.$$

The natural question that arises is: Is there a way to classify conics in this model? For this purpose we study the incidence of the conics with the three additional ideal points. If a conic contains both ideal elements e_3 and e_6 it automatically contains also $e_3 + e_6$. First we look at the entities $\mathfrak{a} \in \bigwedge^1 V$ that contain the ideal points e_3, e_6, and therefore, also $e_3 + e_6$. Thus, we get the conditions

$$\mathfrak{a} \cdot e_3 = 2a_1 = 0, \quad \mathfrak{a} \cdot e_6 = 2a_4 = 0.$$

Hence, a_1 and a_4 have to vanish. The corresponding algebra element has the form

$$\mathfrak{l} = 2a_2 e_2 - a_3 e_3 + 2a_5 e_5 - a_6 e_6.$$

1.10 Quadric Geometric Algebra

Its GIPNS is calculated by

$$\mathbb{NI}_G(\mathfrak{l}) = \left\{ (x,y)^T \in \mathbb{R}^2 \mid 2a_2 x + 2a_5 y + a_3 + a_6 = 0 \right\}.$$

Clearly, this entity describes an inner product line and every line passes through e_3, e_6, and $e_3 + e_6$. An algebra element that contains just e_3 or e_6 is a parabola whose axis are parallel to the x-axis or the y-axis. An element that contains $e_3 + e_6$, but neither e_3 nor e_6, is given by the condition $\mathfrak{a} \cdot (e_3 + e_6) = 2a_1 + 2a_4 = 0$. This means $a_1 = -a_4$ and the corresponding conic is an equilateral hyperbola, i.e., the asymptotes enclose an angle of 90°. All other conics in principal axes position can be obtained by the wedge product of four embedded points or by using the bijection between conics and the algebra elements, cf. Eq. (1.55).

Remark 1.21. *Note, that this description of conics also contains conics without real points.*

In the most general case two-blades correspond to inner product point quadruples. This can be seen from

$$\mathbb{NI}_G(\mathfrak{a} \wedge \mathfrak{b}) = \mathbb{NI}_G(\mathfrak{a}) \cap \mathbb{NI}_G(\mathfrak{b}),$$

see [53]. Therefore, two-blades represent all points belonging to both conics that are represented by the vectors. If two non-degenerate conics do not intersect, the corresponding two-blade represents a complex inner product point quadruple. Furthermore, we can see by dualization, that three-blades correspond to outer product point quadruples. For two inner product lines $\mathfrak{l}_1, \mathfrak{l}_2$ the corresponding two-blade $\mathfrak{l}_1 \wedge \mathfrak{l}_2$ represents an inner product pair of points, where one of the points is their affine intersection point and the other the point at infinity.

Example 1.11. *Let us generate a conic through four points. Therefore, we choose four points and embed them via (1.51).*

$$P_1 = (-1, 0)^T \to \mathfrak{p}_1 = e_1 - e_2 + \frac{1}{2} e_3 + e_4,$$

$$P_2 = (1, 0)^T \to \mathfrak{p}_2 = e_1 + e_2 + \frac{1}{2} e_3 + e_4,$$

$$P_3 = (0, -1)^T \to \mathfrak{p}_3 = e_1 + e_4 - e_5 + \frac{1}{2} e_6,$$

$$P_4 = (-1,0)^{\mathrm{T}} \to \mathfrak{p}_4 = e_1 + e_4 + e_5 + \frac{1}{2}e_6.$$

The corresponding inner product representation is calculated by

$$\mathfrak{c} = \mathfrak{J} \cdot (\mathfrak{p}_1 \wedge \mathfrak{p}_2 \wedge \mathfrak{p}_3 \wedge \mathfrak{p}_4) = 4e_1 - e_3 + 4e_4 - e_6.$$

The GIPNS is given by

$$\mathbb{NI}_G(\mathfrak{c}) = \left\{ (x,y)^T \in \mathbb{R}^2 \mid -x^2 + 1 - y^2 = 0 \right\}.$$

With the bijection (1.55) we see easily that $\mathfrak{c}$ is the image of the conic given by the diagonal matrix $diag(1,-1,-1)$, i.e., the unit circle centered at the origin.

1.10.3 Transformations

In this section we discuss transformations in this algebra. The Clifford algebra $\mathcal{Cl}_{(4,2,0)}$ corresponds to the quadratic space $\mathbb{R}^{(4,2)}$, and therefore, the sandwich action of vectors represents reflections in hyperplanes in this space. It is clear that the transformations induce transformations that are not linear in the base space $\mathbb{R}^2$ because the embedding η is quadratic. From the last section we know that the geometric entity corresponding to vectors are conics in principal position. Furthermore, a transformation acts via the sandwich operator and results again in a k-blade when applied to a k-blade $k = 1, \ldots, 4$. We begin with an example:

Example 1.12. *Let $\mathfrak{c}$ be the circle from Ex. 1.11*

$$\mathfrak{c} = 4e_1 - e_3 + 4e_4 - e_6$$

and let $\mathfrak{p} = e_1 + e_2 + \frac{1}{2}e_3 + e_4 + 2e_5 + 2e_6$ be $\eta(1,2)^{\mathrm{T}}$. Applying the sandwich operator to $\mathfrak{p}$ results in

$$\mathfrak{p}' = \alpha(\mathfrak{c})\mathfrak{p}\mathfrak{c}^{-1} = 5e_1 + e_2 - \frac{1}{2}e_3 + 5e_4 + 2e_5 + e_6.$$

Now we check if this entity is still an embedded point. Therefore, the conditions (1.52) and (1.53) have to be checked. Condition (1.52) is satisfied, but condition (1.53) not

$$\mathfrak{p}'^2_x = 6, \quad \mathfrak{p}'^2_y = -6.$$

1.10 Quadric Geometric Algebra

Hence, $\mathfrak{p}'$ cannot be interpreted as embedded point. The GIPNS of $\mathfrak{p}'$ is given by

$$\mathbb{NI}_G(\mathfrak{p}') = \left\{ (x,y)^T \in \mathbb{R}^2 \mid -\frac{5}{2}x^2 + x - \frac{5}{2}y^2 + 2y - \frac{1}{2} = 0 \right\}.$$

This represents a pair of complex lines intersecting in the real point $(\frac{1}{5}, \frac{2}{5})^T$.

Ex. 1.12 shows that in general a conic is mapped to another conic. We can not map a circle of radius zero to a circle of radius zero and define a mapping for points on this way. Thus, we study the action of the transformations applied to vectors that correspond to conics.

Theorem 1.5. *A conic represented by the vector* $\mathfrak{a} \in \bigwedge^1 V$ *is fixed pointwise under the transformation induced by itself. Furthermore, these transformation are involutions.*

Proof. First, we show that the conic corresponding to the transformation is fixed pointwise. Therefore, we look at the action of a general vector

$$\mathfrak{a} = -2a_1 e_1 + 2a_2 e_2 - a_3 e_3 - 2a_4 e_4 + 2a_5 e_5 - a_6 e_6$$

to itself, and find

$$\alpha(\mathfrak{a})\mathfrak{a}\mathfrak{a}^{-1} = \alpha(\mathfrak{a}) = -\mathfrak{a}.$$

Multiplication with a homogeneous factor does not change the GIPNS. Thus, the result is the conic represented by $\mathfrak{a}$ again. To show that the points of the conic $\mathfrak{a}$ are fixed under the transformation induced by $\mathfrak{a}$, we examine the action of $\mathfrak{a}$ on the intersection points of the conic $\mathfrak{a}$ with all lines containing the point $(0,0)$. These lines are given by

$$\mathfrak{l}(x,y) = \mathfrak{I} \cdot (\eta(0,0) \wedge \eta(x,y) \wedge e_3 \wedge e_6)$$
$$= \mathfrak{I} \cdot \left((e_1 + e_4) \wedge (e_1 + xe_2 + \frac{1}{2}x_1^2 e_3 + e_4 + ye_5 + \frac{1}{2}y^2 e_6) \wedge e_3 \wedge e_6 \right)$$
$$= 2ye_2 - 2xe_5.$$

The intersection of all these lines with the conic $\mathfrak{a}$ is represented by

$$\mathfrak{l}(x,y) \wedge \mathfrak{a} = 4a_1 y e_{12} - 4a_4 y e_{24} - 4a_1 x e_{15} + 4(a_5 y + a_2 x) e_{25}$$
$$- 2a_3 y e_{23} - 2a_3 x e_{35} - 2a_6 y e_{26} - 4a_4 x e_{45} + 2a_6 x e_{56}.$$

This two-blade represents the pair of common points of the conic and $\mathfrak{l}(x,y)$. The application of the transformation induced by $\mathfrak{a}$ to $\mathfrak{l}(x,y) \wedge \mathfrak{a}$ results in

$$\alpha(\mathfrak{a})(\mathfrak{l}(x,y) \wedge \mathfrak{a})\mathfrak{a}^{-1} = 4a_1 y e_{12} - 4a_4 y e_{24} - 4a_1 x e_{15} + 4(a_5 y + a_2 x) e_{25}$$
$$- 2a_3 y e_{23} - 2a_3 x e_{35} - 2a_6 y e_{26} - 4a_4 x e_{45} + 2a_6 x e_{56}.$$

This shows, that all pairs of common points of the pencil of lines with the conic are fixed, and therefore, the whole conic is fixed pointwise. To see that the transformation is an involution we have to apply it twice to an arbitrary k-blade $\mathfrak{B}$

$$\alpha(\mathfrak{a})\alpha(\mathfrak{a})\mathfrak{B}\mathfrak{a}^{-1}\mathfrak{a}^{-1} = \alpha(\mathfrak{a}^2)\mathfrak{B}(\mathfrak{a}^2)^{-1} = \mathfrak{B}.$$

The last equality follows because $\mathfrak{a}^2$ is a real number. $\square$

Due to the fact that these transformations are represented as reflections with respect to hyperplanes in $\mathbb{R}^{(4,2)}$, they are involutions and fix the corresponding hyperplane pointwise. This is the reason why we interpret these transformations as reflections or inversions with respect to conics. Furthermore, the whole group of transformations is generated by the action of vectors. Note that the image of a conic in principal position is always a conic in principal position in this model and that intersection point quadruples of a conic with the reflection conic stay fixed, no matter if the intersection points are real or complex.

Remark 1.22. *The group of conformal transformations of a quadratic space $\mathbb{R}^{(p,q)}$ can be described as the Pin group of a Clifford algebra $\mathcal{C}\ell_{(p+1,q+1,0)}$, see [55]. Therefore, the group of conformal transformations of the Minkowski space $\mathbb{R}^{(3,1)}$ is isomorphic to the group of inversions with respect to conics in principal position except for a translation. Especially for the planar quadric geometric algebra Q2GA we have the signature $(p,q,r) = (4,2,0)$ which is identical to the signature of the homogeneous model where we choose Lie's quadric as metric quadric. Thus, the Pin group of Q2GA is isomorphic to the group of Lie transformations.*

1.10.4 Effect on Lines and Points

In Ex. 1.12 we have seen that a circle with radius zero is mapped to a pair of complex conjugate lines intersecting in a real point. Therefore, we have to search for a better description of points in this model. One way to describe points as two-blades is to examine the intersection of two lines. If we take just the affine point of intersection, we can define an embedding of the affine plane as points of intersection of pairs of lines. Therefore, we take two lines through a given point $(x,y)^T \in \mathbb{R}^2$. We define this point to be the point of intersection of the line parallel to the x-axis and the line parallel to the y-axis.

$$\begin{aligned}\mathfrak{p} &= (\mathfrak{J} \cdot (\eta(x,y) \wedge \eta(x,0) \wedge e_3 \wedge e_6)) \wedge (\mathfrak{J} \cdot (\eta(x,y) \wedge \eta(0,y) \wedge e_3 \wedge e_6)) \\ &= -y^2 x e_{23} - 2yx e_{25} - yx^2 e_{35} + yx^2 e_{56} - y^2 x e_{26} \\ &= 2e_{25} + x(e_{35} - e_{56}) + y(e_{23} + e_{26}). \end{aligned} \quad (1.55)$$

Note that this element represents a pair of points, since lines intersect also in the point ∞. We parametrize affine lines as the sets of all lines passing through two points $p_1 = (x_1, y_1)^T$ and $p_2 = (x_2, y_2)^T$. The inner product line is derived by

$$\begin{aligned}\mathfrak{l}(p_1, p_2) &= \mathfrak{J} \cdot (\eta(p_1) \wedge \eta(p_2) \wedge e_3 \wedge e_6) \\ &= -2(y_1 - y_2)e_2 + (x_1 y_2 - y_1 x_2)e_3 + 2(x_1 - x_2)e_5 + (x_1 y_2 - y_1 x_2)e_6.\end{aligned} \quad (1.56)$$

The GIPNS of this line is determined by

$$\mathbb{NI}_G(\mathfrak{l}(p_1,p_2)) = \{(x,y) \in \mathbb{R}^2 \mid (x_1-x_2)y + (y_2-y_1)x + (y_1 x_2 - x_1 y_2) = 0\}.$$

Theorem 1.6. *The image of a line under an inversion with respect to a conic represented by the non-null vector $\mathfrak{a} \in \bigwedge^1 \mathcal{Cl}_{(4,2,0)}$ is a conic. Moreover, for non-degenerate conics this conic is the image of $\mathfrak{a}$ under an affine transformation, i.e., translation and scalar multiplication.*

Proof. To show this we concentrate on conics with no terms in x and y. We can do this because we are just interested in the type of the image conic. Furthermore, we can perform translations by two reflections in parallel lines, and thus, we can carry over the results from the principal position to an arbitrary position. Furthermore, we just show this theorem for non degenerate conics. A conic with no terms in x or y is given by

$$\mathfrak{a} = 2a_1 e_1 + \frac{1}{2}a_0 e_3 + 2a_4 e_4 + \frac{1}{2}a_0 e_6.$$

Since we are interested in real conics, the coefficients a_0, a_1, and a_4 are not allowed to have the same sign. In the sequel we assume that the conic is real. Now we look at the matrix of the conic

$$M = \begin{pmatrix} 1 & 0 & 0 \\ 0 & \frac{a_1}{a_0} & 0 \\ 0 & 0 & \frac{a_4}{a_0} \end{pmatrix}.$$

The image of the set of lines (1.56) is calculated by

$$\alpha(\mathfrak{a})\mathfrak{l}(p_1, p_2)\mathfrak{a}^{-1} = -\frac{4a_1(x_1 y_2 - y_1 x_2)}{a_0} e_1 - 2(y_1 - y_2)e_2$$
$$- \frac{4a_4(x_1 y_2 - y_1 x_2)}{a_0} e_4 + 2(x_1 - x_2)e_5.$$

The coefficient matrix of the corresponding conic is given by

$$N(p_1, p_2) = \frac{1}{a_0} \begin{pmatrix} 0 & a_0(y_2 - y_1) & a_0(x_1 - x_2) \\ a_0(y_2 - y_1) & 2c_1(x_1 y_2 - y_1 x_2) & 0 \\ a_0(x_1 - x_2) & 0 & 2a_4(x_1 y_2 - y_1 x_2) \end{pmatrix}.$$

From this representation we see immediately that lines through the center of the conic are fixed, but not pointwise. In order to transform this matrix to diagonal form we apply the transformation

$$p \mapsto \underbrace{\begin{pmatrix} 1 & 0 & 0 \\ \alpha & 1 & 0 \\ \beta & 0 & 1 \end{pmatrix}}_{=:T} p, \text{ with } \alpha = -\frac{a_0(y_1 - y_2)}{2a_1(x_1 y_2 - y_1 x_2)} \text{ and } \beta = \frac{a_0(x_1 - x_2)}{2a_4(x_1 y_2 - y_1 x_2)}.$$

Here $p = (1, x, y)^T \in \mathbb{R}^3$ is a point in the projective plane. The action of this coordinate transformation applied to the coefficient matrix of the conic yields

$$N'(p_1, p_2) = T^{-T} N(p_1, p_2) T^{-1}$$

$$= \begin{pmatrix} -\frac{a_0^2(a_4(y_1-y_2)^2+a_1(x_1-x_2)^2)}{2a_4a_1(x_1y_2-y_1x_2)} & 0 & 0 \\ 0 & \frac{2a_1(x_1y_2-y_1x_2)}{a_0} & 0 \\ 0 & 0 & \frac{2a_4(x_1y_2-y_1x_2)}{a_0} \end{pmatrix}.$$

If we look at the affine part of the conic, we see that this results in

$$N'(p_1, p_2) = \begin{pmatrix} 1 & 0 & 0 \\ 0 & ka_1 & 0 \\ 0 & 0 & ka_4 \end{pmatrix}, \text{ with } k = -\frac{4a_1^2(x_1y_2-y_1x_2)^2 a_4}{a_0^2(a_4(y_1-y_2)^2+a_1(x_1-x_2)^2)}.$$

Therefore, the image is identical to the conic corresponding to $\mathfrak{a}$ except for a translation T and a scaling k. Furthermore, lines are mapped to real conics. □

To illustrate this, Fig. 1.2 shows the reflections of three intersecting lines with respect to a circle, an ellipse, a parabola, and a hyperbola. The inversion conic $\mathfrak{a}_i$, $i = 1, \ldots, 4$ is shown in red while the pairs (line, image of the line) are presented in another colour (but the same). The inversion conics are from left to right and from up to down given by

$$\mathfrak{a}_1: x^2 + y^2 = 1, \qquad \mathfrak{a}_2: \frac{25}{16}x^2 + \frac{16}{25}y^2 = 1,$$

$$\mathfrak{a}_3: x^2 - y = 1, \qquad \mathfrak{a}_4: x^2 - \frac{3}{4}y^2 = 1.$$

1.10.5 Subgroups

In this section we examine some subgroups that are embedded naturally in the Pin and the Spin group of Q2GA.

Rotation First, we concentrate on the group that is generated by inversions with respect to lines passing through the origin. Therefore, we study the action of these mappings applied to points embedded via (1.55). Two lines through the origin may be represented by

$$\mathfrak{l}_1 = \mathfrak{J} \cdot (\eta(0,0) \wedge \eta(\cos\varphi, \sin\varphi) \wedge e_3 \wedge e_6),$$
$$\mathfrak{l}_2 = \mathfrak{J} \cdot (\eta(0,0) \wedge \eta(\cos\psi, \sin\psi) \wedge e_3 \wedge e_6).$$

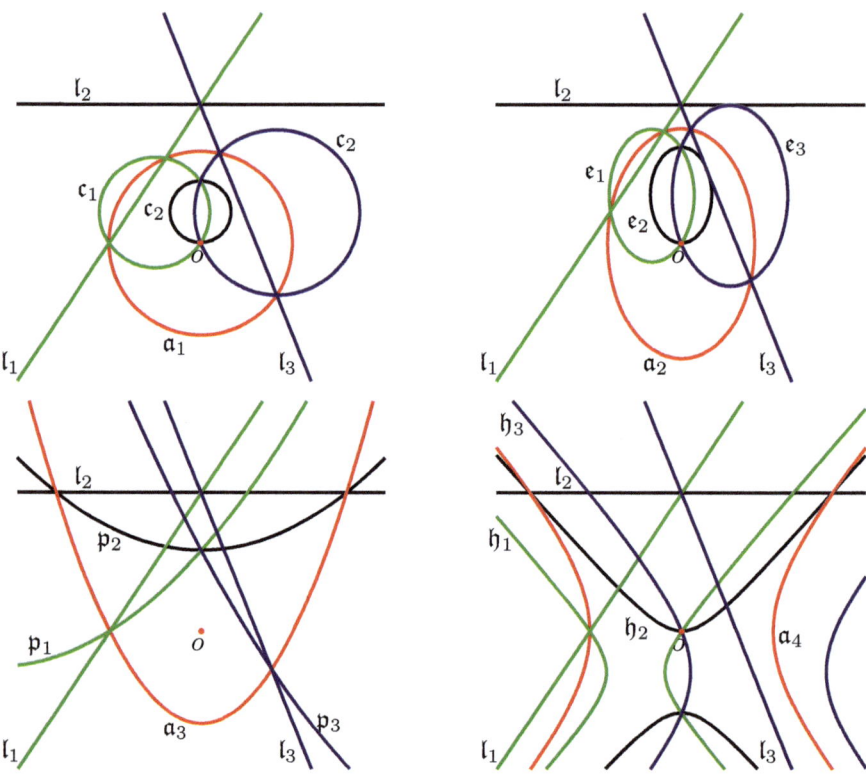

Figure 1.2: Inversions with respect to conics.

Furthermore, we are interested in orientation preserving transformations, *i.e.*, elements from the Spin group. The composition of two reflections in $\mathfrak{l}_1$ and $\mathfrak{l}_2$ is given by their geometric product

$$\mathfrak{l}_1\mathfrak{l}_2 = (\sin\psi\sin\varphi + \cos\psi\cos\varphi) + (\cos\psi\sin\varphi - \sin\psi\cos\varphi)e_{25}$$

and with the addition theorems for sine and cosine we conclude

$$\mathfrak{l}_1\mathfrak{l}_2 = \cos(\varphi-\psi) + \sin(\varphi-\psi)e_{25}. \qquad (1.57)$$

The square of e_{25} is -1. This means that the consecutive reflection in two lines through the origin results in an algebra element that

1.10 Quadric Geometric Algebra

can be interpreted as complex number with norm equal to 1. It is a well-known result that rotations in the plane can be described by normed complex numbers. Let us look at the action of such an element applied to a point that is described by (1.55). Let $\mathfrak{R} = \cos\varphi + \sin\varphi e_{25}$ be an element in the form of Eq. (1.57) and $\mathfrak{p} = 2e_{25} + x_0(e_{35} - e_{56}) + y_0(e_{23} + e_{26})$ a point of the form (1.55). We compute

$$\mathfrak{p}' = \alpha(\mathfrak{R})\mathfrak{p}\mathfrak{R}^{-1} = \mathfrak{R}\mathfrak{p}\mathfrak{R}^{-1}$$
$$= 2e_{25} + (-2\sin\varphi\cos\varphi x_0 + 2\cos(\varphi)^2 y_0 - y_0)(e_{23} + e_{26})$$
$$+ (2\sin\varphi\cos\varphi y_0 + 2\cos(\varphi)^2 x_0 - x_0)(e_{35} - e_{56}).$$

The GIPNS of this entity can be computed or we can simply read the coordinates of the image point.

$$x = 2\cos(\varphi)^2 x_0 + 2\cos\varphi\sin\varphi y_0 - x_0 = \cos(2\varphi)x_0 + \sin(2\varphi)y_0,$$
$$y = 2\cos(\varphi)^2 y_0 - 2\cos\varphi\sin\varphi x_0 - y_0 = \cos(2\varphi)y_0 - \sin(2\varphi)x_0.$$

Therefore, we can see that this transformation, indeed, is a rotation about the origin with the rotation angle 2φ. So these elements constitute a double cover of the group SO(2).

Remark 1.23. *From the fact that vectors are mapped to vectors by a reflection in a line it follows that axis aligned conics have to be mapped to axis aligned conics. Therefore, these mappings can not be interpreted pointwise for conics.*

Translations Now we aim at the group of planar Euclidean displacements SE(2). Therefore, we show that two consecutive reflections in parallel lines result in a translation. The group of planar Euclidean displacements can be generated as the semi-direct product of SO(2) and T(2), which describes the abelian translation group. Let l_1 and l_2 be two parallel lines and let t_1, t_2 be their distances from the origin. The lines are given by

$$l_1(\varphi, t_1) = 2\sin\varphi e_2 - t_1 e_3 - 2\cos\varphi e_5 - t_1 e_6,$$
$$l_2(\varphi, t_2) = 2\sin\phi e_2 - t_2 e_3 - 2\cos\varphi e_5 - t_2 e_6.$$

The composition can be expressed with the geometric product as

$$\mathfrak{T}(\varphi, t_1, t_2) = \mathfrak{l}_1(\varphi, t_1)\mathfrak{l}_2(\varphi, t_2) \quad (1.58)$$
$$= 2 + (t_1 - t_2)\sin\varphi(e_{23} + e_{26}) + (t_1 - t_2)\cos\varphi(e_{35} - e_{56}).$$

Applying the sandwich operator to a point $\mathfrak{p}$ results in

$$\alpha(\mathfrak{T})\mathfrak{p}\mathfrak{T}^{-1} = \mathfrak{T}\mathfrak{p}\mathfrak{T}^{-1}$$
$$= 2e_{25} + (y_0 - 2t_2\cos\varphi + 2t_1\cos\varphi)(e_{23} + e_{26})$$
$$+ (x_0 - 2t_1\sin\varphi + 2t_2\sin\varphi)(e_{35} - e_{56}).$$

The image is determined by

$$x = x_0 - \sin\varphi(2(t_1 - t_2)), \qquad y = y_0 + \cos\varphi(2(t_1 - t_2)).$$

Therefore, the transformation is a translation in the direction normal to the given lines $\mathfrak{l}_1$ and $\mathfrak{l}_2$.

The Group of planar Euclidean Displacements Translations and rotations about the origin generate the entire group of planar Euclidean displacements SE(2). Furthermore, we can now examine the group that is generated by rotations and translations as a subgroup of the Spin group. These algebra elements have the form

$$a_0 + a_1 e_{25} + a_2(e_{23} + e_{26}) + a_3(e_{35} - e_{56}).$$

The multiplication table of the geometric product for the generators e_0, e_{25}, $e_{23} + e_{26}, e_{35} - e_{56}$ is given in Table 1.4. Hence, this is indeed a subgroup of the Spin group. Furthermore, it is isomorphic to a subgroup of the multiplicative group of dual quaternions, called planar dual quaternions. We can define a bijection by

$$e_{25} \mapsto \mathbf{i}, \quad (e_{23} + e_{26}) \mapsto \epsilon\mathbf{j}, \quad (e_{35} - e_{56}) \mapsto \epsilon\mathbf{k}.$$

Remark 1.24. *If we restrict ourself to reflections in lines and circles, we are able to describe the group of conformal transformations of the plane.*

1.10 Quadric Geometric Algebra

Table 1.4: Multiplication table of planar displacements in $\mathcal{C}\ell_{(4,2,0)}$

	1	e_{25}	$e_{23}+e_{26}$	$e_{35}-e_{56}$
1	1	e_{25}	$e_{23}+e_{26}$	$e_{35}-e_{56}$
e_{25}	e_{25}	-1	$e_{35}-e_{56}$	$-(e_{23}+e_{26})$
$e_{23}+e_{26}$	$e_{23}+e_{26}$	$-(e_{35}-e_{56})$	0	0
$e_{35}-e_{56}$	$e_{35}-e_{56}$	$e_{23}+e_{26}$	0	0

Inversions applied to Points In this section we study the action of reflections with respect to a conic in principal position (centered at the origin) on points. The points are embedded as points of intersection of two lines, as discussed in the previous section. Furthermore, we have to note that the transformations map point pairs to point pairs. Hence, the pair of points of intersection of two lines (the affine and the ideal point) are mapped to a pair of points. All lines pass through ∞ and so the image of every line must pass through the image of this point. The generalization to conics that are not in principal position is obtained by the application of a coordinate transformation. The inversion conic is given by

$$\mathfrak{a} = \frac{1}{2}c_0(e_3+e_6) + 2c_1 e_1 + 2c_2 e_4.$$

A point is represented as point of intersection of two lines (see (1.55)) by

$$\mathfrak{p} = 2e_{25} + y_0(e_{23}+e_{26}) + x_0(e_{35}-e_{56}).$$

Applying the sandwich operator to the point results in

$$\mathfrak{p}' = \alpha(\mathfrak{a})\mathfrak{p}\mathfrak{a}^{-1} = -\frac{2y_0 c_1}{c_0}e_{12} + \frac{2x_0 c_1}{c_0}e_{15} - 2e_{25} + \frac{2y_0 c_2}{c_0}e_{24} + \frac{2x_0 c_2}{c_0}e_{45}.$$

This is not of the form (1.55), and therefore, it is not the representation of the intersection of two lines. The GIPNS is calculated

$$\mathbb{NI}_G(\mathfrak{p}') = \{(x,y)^T \in \mathbb{R}^2 \mid -2c_1(xy_0-x_0y)e_1 - (y_0 c_1 x^2 + 2yc_0 + y_0 c_2 y^2)e_2 \\ -2c_2(xy_0-x_0y)e_4 + (x_0 c_1 x^2 + x_0 c_2 y^2 + 2xc_0)e_5 = 0\}$$

The solution set of this GIPNS can be written as

$$x = -\frac{2c_0 x_0}{c_1 x_0^2 + c_2 y_0^2}, \qquad y = -\frac{2c_0 y_0}{c_1 x_0^2 + c_2 y_0^2}.$$

Note that we excluded the solution $x = 0, y = 0$, that is the image of ∞ under the inversion.

Remark 1.25. *Inversion with respect to conics that are not in principal position can be performed by the composition of an inversion and a rotation. Note, that this rotation has to be applied pointwise.*

1.10.6 Generalization to higher Dimensions

The main advantage of this geometric algebra model is its flexibility. It is no problem to change the dimension. We discuss the model for the n-dimensional case and we show some examples for the three-dimensional case. The construction is done in the same way as in section 1.10. We start with a real vector space of dimension n. For each axis we use a conformal embedding. Therefore, the dimension of the geometric algebra is 2^{3n} and its quadratic form is given by

$$Q = \underbrace{\begin{pmatrix} D & & & \\ & D & & \\ & & \ddots & \\ & & & D \end{pmatrix}}_{n\text{-times}}, \quad D = \begin{pmatrix} 0 & 0 & -1 \\ 0 & 1 & 0 \\ -1 & 0 & 0 \end{pmatrix}.$$

The embedding η is realized by

$$\eta : \mathbb{R}^n \to \bigwedge\nolimits^1 V, \tag{1.59}$$

$$(x_1, \ldots, x_n) \mapsto e_1 + x_1 e_2 + \frac{1}{2} x_1^2 e_3 + \ldots + e_{3n-2} + x_n e_{3n-1} + \frac{1}{2} x_n^2 e_{3n}.$$

The conditions for an embedded point (1.52) and (1.53) generalize to

$$\eta(P)^2 = 0, \quad \eta(P)_{x_1}^2 = 0, \quad \eta(P)_{x_2}^2 = 0, \ldots \quad \eta(P)_{x_n}^2 = 0.$$

We define analogue to Def. 1.19:

Definition 1.20. *The blade $\mathfrak{J}$ that maps outer product null spaces to inner product null spaces is defined by*

$$\mathfrak{J} = \bigwedge_{\substack{i=1 \\ i \bmod 3 = 2}}^{n} e_i \wedge \bigwedge_{\substack{j=1 \\ j \bmod 3 = 1}}^{n} e_j \wedge \sum_{\substack{k=1 \\ k \bmod 3 = 0}}^{n} e_k.$$

1.10 Quadric Geometric Algebra

Inner product null spaces can be mapped to outer product null spaces with the blade

$$\mathcal{I}^* := \bigwedge_{\substack{i=1 \\ i \bmod 3 = 2}}^{n} e_i \wedge \bigwedge_{\substack{j=1 \\ j \bmod 3 = 0}}^{n} e_j \wedge \sum_{\substack{k=1 \\ k \bmod 3 = 1}}^{n} e_k.$$

Grade-1 elements correspond to inner product hyperquadrics in principal position. As the dimension is growing, the number of objects that can be represented grows, too. Blades of grade $k, (k \leq n)$ correspond to the intersection of k hyperquadrics.

Quadrics in three-dimensional Space To construct the quadric geometric algebra for the three-dimensional space we use the quadratic space $\mathbb{R}^{(6,3)}$ given by the nine-dimensional real vector space $\mathbb{R}^9$ together with the quadratic form

$$Q = \begin{pmatrix} D & & \\ & D & \\ & & D \end{pmatrix}, \quad D = \begin{pmatrix} 0 & 0 & -1 \\ 0 & 1 & 0 \\ -1 & 0 & 0 \end{pmatrix}.$$

For three dimensions the embedding η, see Eq (1.59), has the following form

$$\eta : \mathbb{R}^3 \to \bigwedge^1 V,$$

$$(x, y, z)^T \mapsto e_1 + xe_2 + \frac{1}{2}x^2 e_3 + e_4 + ye_5 + \frac{1}{2}y^2 e_6 + e_7 + ze_8 + \frac{1}{2}z^2 e_9.$$

The conditions for an embedded point (1.52) and (1.53) generalize to

$$\eta(P)^2 = 0, \quad \eta(P)_x^2 = 0, \quad \eta(P)_y^2 = 0, \quad \eta(P)_z^2 = 0.$$

Moreover, the blades $\mathcal{I}$ and $\mathcal{I}^*$ are given by

$$\mathcal{I} = (e_2 \wedge e_5 \wedge e_8) \wedge (e_1 \wedge e_4 \wedge e_7) \wedge (e_3 + e_6 + e_9),$$
$$\mathcal{I}^* = (e_2 \wedge e_5 \wedge e_8) \wedge (e_3 \wedge e_6 \wedge e_9) \wedge (e_1 + e_4 + e_7).$$

The corresponding geometric algebra has dimension $2^9 = 512$. Any quadric in principal position except for translation in $\mathbb{R}^3$ is uniquely determined by six values. This can be seen from the symmetric matrix of the equation of the quadric that has, in general, ten free entries. The fact that we are treating quadrics in principal position reduces the number of free entries to seven. Furthermore, this matrix representation is homogeneous, and therefore, we have six degrees of freedom. In analogy to Eq. (1.55) we obtain a bijection χ that is defined as

$$C \mapsto 2c_1e_1 - 2c_2e_2 + \frac{1}{3}c_0e_3 + 2c_4e_4 - 2c_5e_5 + \frac{1}{3}c_0e_6 + 2c_7e_7 - 2c_8e_8 + \frac{1}{3}c_0e_9,$$

with

$$C = \begin{pmatrix} c_0 & c_2 & c_5 & c_8 \\ c_2 & c_1 & 0 & 0 \\ c_5 & 0 & c_4 & 0 \\ c_8 & 0 & 0 & c_7 \end{pmatrix}.$$

As we did for the planar case, we can now pay our attention on the intersection of three planes in order to get a pair of points containing one affine and one ideal point. We choose these planes to be parallel to the coordinate planes and passing through a given point $P = (x_0, y_0, z_0)^T$. Expressed in terms of the quadric geometric algebra $\mathcal{C}\ell_{(6,3,0)}$ we get

$$\begin{aligned}
\mathfrak{p} &= (\eta(x_0, y_0, z_0)^T \wedge \eta(0, 0, z_0)^T \wedge \eta(x_0, 0, z_0)^T \wedge e_3 \wedge e_6 \wedge e_9) \cdot \mathfrak{J} \\
&\wedge (\eta(x_0, y_0, z_0)^T \wedge \eta(0, y_0, 0)^T \wedge \eta(x_0, y_0, 0)^T \wedge e_3 \wedge e_6 \wedge e_9) \cdot \mathfrak{J} \\
&\wedge (\eta(x_0, y_0, z_0)^T \wedge \eta(x_0, 0, 0)^T \wedge \eta(x_0, y_0, 0)^T \wedge e_3 \wedge e_6 \wedge e_9) \cdot \mathfrak{J} \\
&= 3e_{258} + (e_{358} - e_{568} + e_{589})x_0 + (e_{238} + e_{268} - e_{289})y_0 \\
&\quad + (-e_{235} + e_{256} + e_{259})z_0.
\end{aligned}$$

Remark 1.26. *A representation of the group of Euclidean displacements* SE(3) *can be obtained by studying the composition of reflections in planes. Planes correspond to vectors that are obtained by* $(\eta P_1 \wedge \eta P_2 \wedge \eta P_3 \wedge e_3 \wedge e_6 \wedge e_9) \cdot \mathfrak{J}$.

Now we define an inner product inversion quadric with

$$P_1 = (\frac{9}{10}, 0, 0)^T, \qquad P_2 = (-\frac{9}{10}, 0, 0)^T, \qquad P_3 = (0, \frac{3}{4}, 0)^T$$

$$P_4 = (0, -\frac{3}{4}, 0)^T, \qquad P_5 = (0, 0, \frac{5}{4})^T, \qquad P_6 = (0, 0, -\frac{5}{4})^T$$

1.10 Quadric Geometric Algebra

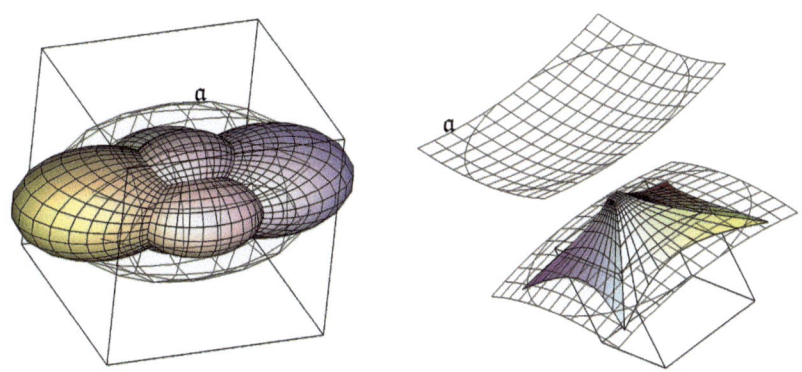

Figure 1.3: Inversion of the unit cube with respect to an ellipsoid (left), inversion with respect to a hyperboloid (right).

$$\mathfrak{a} = (\eta P_1 \wedge \eta P_2 \wedge \eta P_3 \wedge \eta P_4 \wedge \eta P_5 \wedge \eta P_6) \cdot \mathfrak{J}$$
$$= \frac{25}{9}e_1 - \frac{3}{8}e_3 + 4e_4 - \frac{3}{8}e_6 + \frac{36}{25}e_7 - \frac{3}{8}e_9.$$

The GIPNS of $\mathfrak{a}$ is given by

$$\mathbb{NI}_G(\mathfrak{a}) = \left\{ (x, y, z)^{\mathrm{T}} \in \mathbb{R}^3 \,\middle|\, \frac{100}{81}x^2 + \frac{16}{9}y^2 + \frac{16}{25}z^2 - 1 = 0 \right\}.$$

In Fig. 1.3 (left) we can see the inversion of the unit cube $[-1, 1]^3$ with respect to the ellipsoid defined by $\mathfrak{a}$. The image of every face of the cube, i.e., of every plane is an ellipsoid passing through the origin. The action of the inversion given by $\mathfrak{a}$ on pairs of points can be written as

$$f(x, y, z) = \frac{45^2}{2^2(30^2 y^2 + 25^2 x^2 + 18^2 z^2)} (x, y, z)^{\mathrm{T}}.$$

Note that the point ∞ is mapped to the origin and that the origin is mapped to ∞. Therefore, we have a map from $\mathbb{R}^3 \setminus \{0\} \to \mathbb{R}^3$. One main advantage of this method is that we can calculate the image ellipsoid of one face of the cube (one plane) directly by applying the sandwich operator to the plane that is expressed as quadric. For example the plane passing through $P_1 = (1,1,1)^T$, $P_2 = (1,1,-1)^T$, $P_3 = (1,-1,1)^T$ can be expressed as vector by

$$\mathfrak{p} = (\eta P_1 \wedge \eta P_2 \wedge \eta P_3 \wedge e_3 \wedge e_6 \wedge e_9) \cdot \mathfrak{I} = 3e_2 + e_3 + e_6 + e_9.$$

To check that that this indeed the representation of the plane we compute the GIPNS

$$\mathbb{NI}_G(\mathfrak{p}) = \{(x,y,z)^T \in \mathbb{R}^3 \mid 1 - x = 0\}.$$

Applying the sandwich operator to $\mathfrak{p}$ results in

$$\mathfrak{c} = \alpha(\mathfrak{a})\mathfrak{p}\mathfrak{a}^{-1} = -\frac{200}{27}e_1 - 3e_2 - \frac{32}{3}e_4 - \frac{96}{25}e_7$$

with GIPNS

$$\mathbb{NI}_G(\mathfrak{c}) = \left\{(x,y,z)^T \in \mathbb{R}^3 \;\middle|\; \frac{10^2}{9^2}x^2 - x + \frac{4^2}{3^2}y^2 + \frac{4^2}{5^2}z^2 = 0\right\}.$$

This is one of the ellipsoids displayed in Fig. 1.3 (left). Furthermore, we can intersect two inner product planes $\mathfrak{p}_1, \mathfrak{p}_2$ to get an inner product line that is an edge of the cube. After that we can apply the sandwich operator to the line and get the intersection curve (an ellipse) of the two ellipsoids that are the images of $\mathfrak{p}_1, \mathfrak{p}_2$.

Remark 1.27. *It is more convenient to compute the sandwich operator by a conjugation instead inversion. This means we use $\alpha(\mathfrak{a})\mathfrak{X}\mathfrak{a}^*$ with $\mathfrak{a} \in \mathcal{C}\ell^\times_{(p,q,r)}$ and $\mathfrak{X} \in \bigwedge^k V$. In general, the modified sandwich operator is easier to handle, because computing the inverse of a Clifford algebra element is extremely expensive. Moreover, we are working in a projective setting, and therefore, multiplication with a homogeneous factor does not change the occurring geometric inner product and outer product null spaces.*

The second example in three-dimensional space is a hyperboloid of two sheets in principal position that is generated by

$$\mathfrak{a} = (\eta P_1 \wedge \eta P_2 \wedge \eta P_3 \wedge \eta P_4 \wedge \eta P_5 \wedge \eta P_6) \cdot \mathfrak{I} = -6e_1 + e_3 + \frac{9}{8}e_4 + e_6 + \frac{9}{8}e_7 + e_9.$$

1.10 Quadric Geometric Algebra

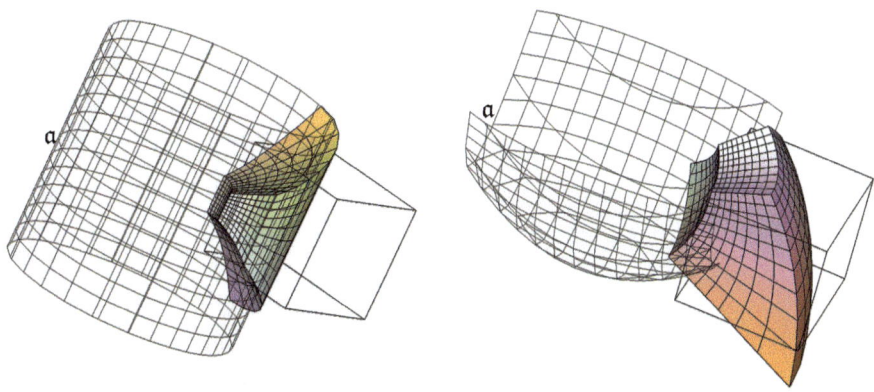

Figure 1.4: Inversion with respect to a cylinder (left), inversion with respect to an elliptic paraboloid (right).

Here, we have $P_1 = (-1,0,0)^\text{T}$, $P_2 = (1,0,0)^\text{T}$, $P_3 = (2,0,4)^\text{T}$, $P_4 = (2,0,-4)^\text{T}$, $P_5 = (2,4,0)^\text{T}$, $P_6 = (2,-4,0)^\text{T}$. We calculate the GIPNS

$$\mathbb{NI}_G(\mathfrak{a}) = \left\{ (x,y,z)^\text{T} \in \mathbb{R}^3 \mid x^2 - 1 - \frac{3}{16}y^2 - \frac{3}{16}z^2 = 0 \right\}.$$

The mapping applied to pairs of points results in

$$f(x,y,z) = \frac{16}{16x^2 - 3y^2 - 3z^2}(x,y,z)^\text{T}.$$

The image of the cube $[1,3] \times [-1,1] \times [-1,1]$ under this mapping is shown in Fig. 1.3 (right). Fig. 1.4 (left) shows the image of an inversion with respect to a cylinder given by $\mathfrak{a} = 3e_1 - 2e_3 + 3e_4 - 2e_6 - 2e_9$ applied to a cube. The equation of the cylinder is derived as $x^2 + y^2 = 4$. Planes that are not parallel to the axis of the cylinder are mapped to paraboloids. Another example is presented in Fig. 1.4 (right). The inversion quadric is an elliptic paraboloid given by $\mathfrak{a} = 3e_1 - 2e_3 + 12e_4 - 2e_6 + 6e_8 - 2e_9$ respectively by $x^2 + y^2 - 4z + 4 = 0$.

2 Chain Geometry over Clifford Algebras

In the following chapter we give an introduction to chain geometry. It is not the aim of this work to give a complete treatise of this topic, we just introduce the concepts we need for our purposes. For a more detailed introduction the reader is referred to [11]. The roots of chain geometry can be found in BENZ [5]. BENZ investigated projective lines over commutative two-dimensional algebras and the corresponding chain geometries. A more recent treatise is [33].

To provide the preliminaries we follow [11] in order to define the setting that is necessary for our purposes. After this introduction to chain geometry and chain geometric concepts we focus on chain geometry over Clifford algebras. We introduce chain geometries over Pin and Spin groups. Moreover, we show that the connected components of the Pin and the Spin group define sub chain geometries of the Clifford algebra. Furthermore, we show that subgroups of these groups also define sub chain geometries. For application to kinematics we study the chain geometry over the Clifford algebra $\mathcal{C}\ell_{(3,0,1)}$ and its Spin group in detail and classify the occurring chains.

2.1 Chain Geometry

2.1.1 Distance Spaces

In this section we recall the concept of a distance space and give some examples that we use later on. Projective lines over rings serve as examples for distance spaces.

Definition 2.1. *A pair* $(\mathcal{P}, \Delta)$, *where* $\mathcal{P}$ *is a non empty set and* $\Delta \subseteq \mathcal{P} \times \mathcal{P}$ *is a relation on* $\mathcal{P}$ *is called a* distance space *if the following conditions are satisfied:*

(1) Δ *is a* symmetric *relation, i.e.,* $a\Delta b \Rightarrow b\Delta a$,

(2) Δ *is an* anti-reflexive *relation, i.e.,* $\neg(a\Delta a)$ *is valid for* $a \in \mathcal{P}$.

The elements of $\mathcal{P}$ *are called points and the relation* Δ *is the* distance *relation. Two points* $a, b \in \mathcal{P}$ *are called distant if* $a\Delta b$.

Definition 2.2. *Elements* a, b *that are not distant are called* parallel $(a \parallel b)$. *The parallel relation is symmetric and reflexive.*

Examples for distance spaces will be given in the following sections.

2.1.2 The projective Line over an $\mathcal{L}$-algebra

First, we give the definition of an $\mathcal{L}$-algebra.

Definition 2.3. *Let* $\mathcal{L}$ *be a subring of* $\mathcal{R}$ *contained in the* center $\mathcal{C}(\mathcal{R})$, *where*
$$\mathcal{C}(\mathcal{R}) := \{a \in \mathcal{R} \mid ax = xa \text{ for all } x \in \mathcal{R}\}.$$
We call $\mathcal{R}$ *an* $\mathcal{L}$-algebra. *An* $\mathcal{L}$-algebra $\mathcal{R}$ *is a module over the commutative ring* $\mathcal{L}$.

Remark 2.1. *In the most cases we deal with algebras over the real numbers. In this case we have* $\mathcal{L} = \mathbb{R}$ *and we call the algebra an* $\mathbb{R}$-algebra.

Definition 2.4. *Let* $\mathcal{R}$ *be a* $\mathcal{K}$-algebra *over the field* $\mathcal{K}$.

(1) For $x \in \mathcal{R}$ *we define* $e(x) := \{a \in \mathcal{K} \mid x + a \in \mathcal{R}^\times\}$.

(2) The $\mathcal{K}$-algebra $\mathcal{R}$ *is called* strong, *if for all* $x \in \mathcal{R}$ *the inequality* $\operatorname{card}(e(x)) > \operatorname{card}(\mathcal{K} \setminus e(x))$ *holds.*

Definition 2.5. *The* general linear group $\operatorname{GL}(\mathcal{R}, 2)$ *over a ring* $\mathcal{R}$ *is defined as the set of invertible* 2×2 *matrices with entries in* $\mathcal{R}$.

With the general linear group we can now define the projective line over a ring $\mathcal{R}$.

2.1 Chain Geometry

Definition 2.6. *The projective line over a ring $\mathcal{R}$ is defined as the set $\mathbb{P}^1(\mathcal{R})$ of all cyclic submodules $\mathcal{R}(a,b)$ of $\mathcal{R}^2$, where (a,b) is the first row of an invertible 2×2 matrix over $\mathcal{R}$:*

$$\mathbb{P}^1(\mathcal{R}) := \left\{ \mathcal{R}(a,b) \mid \exists c, d : \begin{pmatrix} a & b \\ c & d \end{pmatrix} \in \mathrm{GL}(\mathcal{R}, 2) \right\}.$$

Such pairs are called admissible, *see [32]. Note, that we define points of the projective line as left-homogeneous equivalence classes.*

Furthermore, we define equivalence classes of admissible pairs, respectively points.

Definition 2.7. *Two points $(a,b), (c,d) \in \mathbb{P}^1(\mathcal{R})$ are equivalent if there is a unit $r \in \mathcal{R}^\times$ with $ra = c$ and $rb = d$. This relation is an equivalence relation and denoted by $\sim$. For an equivalence class we write $\mathcal{R}(a,b)$.*

The homogeneous component of the pair is the second coordinate. This means the algebra element 0 has the projective coordinates $U = \mathcal{R}(0,1)$, the neutral element 1 has the coordinates $V = \mathcal{R}(1,1)$, and the projective point $W = \mathcal{R}(1,0)$ corresponds to an ideal point. In the following we denote points of the projective line $\mathbb{P}^1(\mathcal{R})$ with capital latin letters.

Remark 2.2. *If $\mathcal{R}$ is a field, Def. 2.6 describes the classical projective line.*

From Def. 2.6 we see, $\mathcal{R}(1,0)$ is a point, since the 2×2 identity matrix is in $\mathrm{GL}(\mathcal{R}, 2)$. If we take a general matrix $\mathrm{M} \in \mathrm{GL}(\mathcal{R}, 2)$ all points $P \in \mathbb{P}^1(\mathcal{R})$ can be obtained by $\mathcal{R}(1,0) \cdot \mathrm{M}$. Hence, the point set $\mathbb{P}^1(\mathcal{R})$ can also be described as the orbit of $\mathcal{R}(1,0)$ under the action of $\mathrm{GL}(\mathcal{R}, 2)$. Two points $X = \mathcal{R}(x_1, x_0), Y = \mathcal{R}(y_1, y_0)$ are called distant $(X \Delta Y)$ if they are complementary, this means if $\mathcal{R}^2 = X \oplus Y$. Hence, $(\mathbb{P}^1(\mathcal{R}), \Delta)$ is a distance space called the projective line over $\mathcal{R}$.

Remark 2.3. *We consider only algebras where the left-inverse elements are also right-inverse elements. Thus, every point may be represented by an admissible pair, see [32].*

2.1.3 The Projective Linear Group $\mathrm{PGL}(\mathcal{R}, 2)$

The group of all linear mappings that map the projective line $\mathbb{P}^1(\mathcal{R})$ onto itself can be described by the general linear group over $\mathcal{R}$. The kernel of $\mathrm{GL}(\mathcal{R}, 2)$ is given by

$$\ker \mathrm{GL}(\mathcal{R}, 2) = \left\{ \begin{pmatrix} q & 0 \\ 0 & q \end{pmatrix} \mid q \in \mathcal{C}(\mathcal{R}) \cap \mathcal{R}^\times \right\}.$$

Note, that the kernel is equal to the center of the general linear group

$$\ker \mathrm{GL}(\mathcal{R}, 2) = \mathcal{C}(\mathrm{GL}(\mathcal{R}, 2)).$$

The *projective linear group* $\mathrm{PGL}(\mathcal{R}, 2)$ acting on $\mathbb{P}^1(\mathcal{R})$ can be defined as the quotient group

$$\mathrm{PGL}(\mathcal{R}, 2) := \mathrm{GL}(\mathcal{R}, 2)/\mathcal{C}(\mathrm{GL}(\mathcal{R}, 2)).$$

Due to the fact, that the projective line is defined as left-module, a mapping of $\mathbb{P}^1(\mathcal{R})$ onto itself can be described with the matrix vector product $(a, b) \cdot M$, where the admissible pair (a, b) is a row vector.

Remark 2.4. *In this chapter we denote the action of an element $\gamma \in \mathrm{PGL}(\mathcal{R}, 2)$ as superscript. For example the image of a point $P \in \mathbb{P}^1(\mathcal{R})$ under γ is denoted by P^γ.*

Theorem 2.1. *The group $\mathrm{PGL}(\mathcal{R}, 2)$ acts on $\mathbb{P}^1(\mathcal{R})$ and leaves the distance relation invariant. This means for $\gamma \in \mathrm{PGL}(\mathcal{R}, 2)$ and $P, Q \in \mathbb{P}^1(\mathcal{R})$ we have $P \Delta Q \Rightarrow P^\gamma \Delta Q^\gamma$, where P^γ and Q^γ denote the images of P and Q under γ.*

A proof of this theorem can be found in [11]. Now we give another theorem that describes the action of $\mathrm{PGL}(\mathcal{R}, 2)$ on $\mathbb{P}^1(\mathcal{R})$ more detailed. Therefore, we need:

Definition 2.8. *Let $(\mathcal{P}, \Delta)$ be a distance space and Γ a subgroup of the group $\mathrm{Aut}(\mathcal{P}, \Delta)$, i.e., the group of automorphisms of the distance space $(\mathcal{P}, \Delta)$. The action of Γ on $(\mathcal{P}, \Delta)$ is called*

(1) 2-Δ-transitive, if Γ acts transitive on the set of all pairs of distant points of $\mathcal{P}$,

2.1 Chain Geometry

(2) 3-Δ-transitive, if Γ acts transitive on the set of all triples of distant points of $\mathcal{P}$.

Remark 2.5. *A group acts transitive on a set if for each pair of elements a,b there is a group element that maps a to b.*

Theorem 2.2. *The group $\mathrm{PGL}(\mathcal{R},2)$ acts 3-Δ-transitive on $\mathbb{P}^1(\mathcal{R})$.*

For a proof we refer to [33].

Table 2.1: Elements of $\mathrm{PGL}(\mathcal{R},2)$ corresponding to addition, multiplication, and reciprocation.

addition	right-multiplic.	left-multiplic.	reciprocation
$x \mapsto x+t$	$x \mapsto xq$	$x \mapsto qx$	$x \mapsto x^{-1}$
$\mathcal{R}(x,1)\begin{pmatrix} 1 & 0 \\ t & 1 \end{pmatrix}$	$\mathcal{R}(x,1)\begin{pmatrix} q & 0 \\ 0 & 1 \end{pmatrix}$	$\mathcal{R}(x,1)\begin{pmatrix} 1 & 0 \\ 0 & q \end{pmatrix}$	$\mathcal{R}(x,1)\begin{pmatrix} 0 & 1 \\ 1 & 0 \end{pmatrix}$

Now we want to investigate a subgroup of the projective group that corresponds to special operations on the ring $\mathcal{R}$. Let $\mathcal{R}(x_1,x_0)$ be an arbitrary point on $\mathbb{P}^1(\mathcal{R})$ then addition, multiplication, and reciprocation in $\mathcal{R}$ can be described by the elements of $\mathrm{PGL}(\mathcal{R},2)$ listed in Table 2.1. These are all permutations of $\mathbb{P}^1(\mathcal{R})$ for that the inverse map is of the same type. The subgroup of $\mathrm{PGL}(\mathcal{R},2)$ generated by these four elements is denoted by $\Gamma(\mathcal{R})$.

2.1.4 The projective Line over a Subring

For a chain geometry we need one more ingredient. Therefore, we look at pairs of rings $(\mathcal{L},\mathcal{R})$ where $\mathcal{L}$ is a subring of $\mathcal{R}$. The projective line over $\mathcal{L}$ is a distance space $(\mathbb{P}^1(\mathcal{L}),\Delta_{\mathcal{L}})$. Hence, we are interested in the relation between $(\mathbb{P}^1(\mathcal{L}),\Delta_{\mathcal{L}})$ and $(\mathbb{P}^1(\mathcal{R}),\Delta_{\mathcal{R}})$. We interpret the projective line over $\mathcal{L}$ as a substructure of $(\mathbb{P}^1(\mathcal{R}),\Delta_{\mathcal{R}})$. It is clear that for a pair $(a,b) \in \mathcal{L}^2$ the equivalence classes have the form $\mathcal{L}(a,b)$ and if we interpret the point as element of $(\mathbb{P}^1(\mathcal{R}),\Delta_{\mathcal{R}})$, the equivalence classes have the form $\mathcal{R}(a,b)$. Hence, we can define the map

$$\iota: \mathbb{P}^1(\mathcal{L}) \to \mathbb{P}^1(\mathcal{R}), \quad \mathcal{L}(a,b) \mapsto \mathcal{R}(a,b).$$

This map is an injective morphism of distance spaces, see [11].

Remark 2.6. *In general we can not expect that non-distant points are mapped to non-distant points under ι. As example we can take a look at the ring pair $(\mathbb{Z}, \mathbb{Q})$, cf. [11, p. 26].*

If we equip $\mathbb{P}^1(\mathcal{L})^\iota$ with the distance relation $\Delta_\mathcal{R}$ the inverse map ι^{-1} does not need to be a morphism of distance spaces. A criteria for the inverse map ι^{-1} to be a morphism of distance spaces is given in:

Theorem 2.3. *Let $\mathcal{L} \leq \mathcal{R}$ and let $\iota: (\mathbb{P}^1(\mathcal{L}), \Delta_\mathcal{L}) \to (\mathbb{P}^1(\mathcal{R}), \Delta_\mathcal{R})$ be a morphism of distance spaces. The following statements are equivalent:*

(1) $\mathcal{L}^\times = \mathcal{R}^\times \cap \mathcal{L}$.

(2) ι maps non-distant points to non-distant points.

In this case, the map $\iota: \mathbb{P}^1(\mathcal{L}) \to \mathbb{P}^1(\mathcal{L})^\iota$ is a isomorphism of distance spaces.

For a proof we refer to [11].

2.2 Chain Geometry as Incidence Geometry

In this section we define the term chain geometry. Therefore, we need the projective line $\mathbb{P}^1(\mathcal{R})$ over an algebra as the point set $\mathcal{P}$ of our chain geometry. Furthermore, there are special subsets of $\mathcal{P}$ that we call chains.

Definition 2.9. *Let $\mathcal{R}$ be an $\mathcal{L}$-algebra and $\gamma \in \mathrm{PGL}(\mathcal{R}, 2)$. The subset $\mathbb{P}^1(\mathcal{L})^\gamma$ of $\mathbb{P}^1(\mathcal{R})$ is called a chain in $\mathbb{P}^1(\mathcal{R})$, if γ is induced by a matrix $\begin{pmatrix} a & b \\ c & d \end{pmatrix} \in \mathrm{GL}(\mathcal{R}, 2)$. The set of all chains in $\mathbb{P}^1(\mathcal{R})$ is denoted by $\mathfrak{C}(\mathcal{L}, \mathcal{R})$ and defined as*

$$\mathfrak{C}(\mathcal{L}, \mathcal{R}) := \{\mathbb{P}^1(\mathcal{L})^\gamma \mid \gamma \in \mathrm{PGL}(\mathcal{R}, 2)\}.$$

We call the incidence structure $\Sigma(\mathcal{L}, \mathcal{R}) := (\mathbb{P}^1(\mathcal{R}), \mathfrak{C}(\mathcal{L}, \mathcal{R}))$ chain geometry over the $\mathcal{L}$-algebra $\mathcal{R}$.

2.2 Chain Geometry as Incidence Geometry

The set of points of $\Sigma(\mathcal{L},\mathcal{R})$ is additionally equipped with the distance relation $\Delta_\mathcal{R}$. We need that the distance relation $\Delta_\mathcal{R}$ on the chain $\mathbb{P}^1(\mathcal{L})$ coincides with $\Delta_\mathcal{L}$. With Th. 2.3 this is the case if $\mathcal{L}^\times = \mathcal{R}^\times \cap \mathcal{L}$. Hence, we assume this from now on and for $\Delta_\mathcal{R}$ we simply write Δ. Furthermore, we provide the more general definition of a chain space that also can be found in [11].

Definition 2.10. *Let $(\mathcal{P},\mathfrak{C})$ be an incidence structure and Δ a relation on $\mathcal{P}$ such that $(\mathcal{P},\Delta)$ is a distance space.*

(1) Then $(\mathcal{P},\mathfrak{C},\Delta)$ is called incidence structure with distance relation.

(2) Let $(\mathcal{P},\mathfrak{C},\Delta)$ be an incidence structure with distance relation. The elements of $\mathfrak{C}$ are called chains. $(\mathcal{P},\mathfrak{C},\Delta)$ is called a weak chain space, *if the following axioms hold:*

C1 *Each chain $c \in \mathfrak{C}$ contains at least three points and each point $p \in \mathcal{P}$ is contained by at least one chain.*

C2 *Three pairwise distant points $p, q, r \in \mathcal{P}$ are incident with exactly one chain $c \in \mathfrak{C}$. We denote this chain by $c =: (pqr)$.*

A weak chain space $(\mathcal{P},\mathfrak{C},\Delta)$ is called a weak chain space in the proper sense if the following additional axiom is satisfied.

C3 *Two points $p, q \in \mathcal{P}$ are distant if, and only if, they are different and connected by a chain.*

We need a further definition.

Definition 2.11. *Let $\mathbb{A} = (P, \mathfrak{L}, \|)$ be an affine space and let $\mathfrak{L}' \subseteq \mathfrak{L}$ be a non-empty union of sets of parallel lines from $\mathbb{A}$. The incidence structure $(P, \mathfrak{L}')$ is called a* partial affine space.

With Def. 2.10 and Def. 2.11, we are able to define chain spaces.

Definition 2.12. *Let $\Sigma = (\mathcal{P},\mathfrak{C},\Delta)$ be a weak chain space in the proper sense.*

(1) For $P \in \mathcal{P}$ let $\mathfrak{C}_P := \{c \setminus \{P\} \mid P \in c \in \mathfrak{C}\}$. Let $\Delta(P)$ denote all points distant from $P \in \mathcal{P}$. Than $\Sigma_P := (\Delta(P), \mathfrak{C}_P)$ is called the residuum *of Σ at the point P.*

(2) If Σ satisfies the axiom

C4 For each $P \in \mathcal{P}$, the residuum Σ_P is a partial affine space.

Σ is called a chain space.

Remark 2.7. *All chain geometries over algebras are also chain spaces, see [11] or [33]. When we talk about chain spaces we mention chain geometries over algebras. A chain geometry over an $\mathcal{L}$-algebra $\mathcal{R}$ is denoted by $\Sigma(\mathcal{L}, \mathcal{R})$, where the set of points is given by $\mathcal{P} = \mathbb{P}^1(\mathcal{R})$ and the set of chains by*

$$\mathfrak{C}(\mathcal{L}, \mathcal{R}) := \{(\mathbb{P}^1(\mathcal{L}))^\gamma | \gamma \in \mathrm{PGL}(\mathcal{R}, 2)\}.$$

In the following we cite some theorems concerning chain geometries over algebras without proofs. For proofs we refer to [11] and [33].

Theorem 2.4. *Let $\mathcal{R}$ be an $\mathcal{L}$-algebra with $\mathcal{L}^\times = \mathcal{R}^\times \cap \mathcal{L}$. Then for $\Sigma(\mathcal{L}, \mathcal{R})$ it is true: Through three pairwise distant points there is exactly one chain.*

Theorem 2.5. *For $\Sigma(\mathcal{L}, \mathcal{R})$ it is equivalent:*

(1) Let $\mathcal{L}^\times = \mathcal{R}^\times \cap \mathcal{L}$. Two pairwise distant points of $\mathbb{P}^1(\mathcal{R})$ are distant if, and only if, they are incident with one chain.

(2) The ring $\mathcal{L}$ is a field.

Lemma 2.1. *Let $\mathcal{R}$ be an algebra over the ring $\mathcal{L}$. Then the group $\mathrm{PGL}(\mathcal{R}, 2)$ is a subgroup of the group of automorphisms $\mathrm{Aut}(\Sigma(\mathcal{L}, \mathcal{R}))$ of the incidence structure $\Sigma = \Sigma(\mathcal{L}, \mathcal{R})$. The group acts transitive on the set $\mathfrak{F} = \{(p, c) \in (\mathcal{P} \times \mathfrak{C}) \mid p \in c \in \mathfrak{C}(\mathcal{L}, \mathcal{R})\}$ of flags of Σ.*

Proposition 2.1. *The parallel relation in $\Sigma(\mathcal{L}, \mathcal{R})$ on $\mathbb{P}^1(\mathcal{R})$ is transitive, i.e., it is an equivalence relation, if $\mathcal{R}$ is a local ring.*

A proof can be found in [33].

Definition 2.13. *Let $\mathcal{K}$ be a field.*

(1) The chain geometry $\Sigma(\mathcal{K}, \mathcal{R})$ is called a Möbius geometry if the parallel relation is the equality relation. This means that any two different points are incident with a single chain.

2.2 Chain Geometry as Incidence Geometry

(2) $\Sigma(\mathcal{K}, \mathcal{R})$ *is called a* Laguerre geometry *provided that the parallel relation is an equivalence relation on* $\mathbb{P}^1(\mathcal{R})$ *and every chain meets every parallel class of points.*

(3) $\Sigma(\mathcal{K}, \mathcal{R})$ *is called a* Minkowski geometry *of dimension n if*

$$\mathcal{R} = \mathcal{K} \times \mathcal{K} \times \ldots \times \mathcal{K} \ (n\text{-times}),$$

where addition and multiplication are defined component-wise.

Remark 2.8. *The chain geometry* $\Sigma(\mathcal{K}, \mathcal{R})$ *is a Laguerre geometry if* $\mathcal{R}$ *is a Laguerre algebra.*

2.2.1 Definition of a Cross Ratio

For a better description of chains and as natural invariant we define the cross ratio of four points.

Definition 2.14. *For* $A, B, C, D \in \mathbb{P}^1(\mathcal{R})$*, where* A, B, C *are mutually distant and* A, D *are distant, we define the* cross ratio $cr(A, B, C, D)$ *as a subset of* $\mathcal{R}$ *in the following way:*

$$d \in cr(A, B, C, D) \Leftrightarrow \exists \gamma \in \mathrm{PGL}(\mathcal{R}, 2) \colon (A^\gamma, B^\gamma, C^\gamma, D^\gamma) = (U, V, W, \mathcal{R}(d, 1)),$$

with $U = \mathcal{R}(0, 1)$, $V = \mathcal{R}(1, 1)$, $W = \mathcal{R}(1, 0)$.

Every cross ratio is a class of conjugates under $\mathcal{R}^\times$:

$$d \in cr(A, B, C, D) \Leftrightarrow cr(A, B, C, D) = \{z^{-1} d z : z \in \mathcal{R}^\times\}.$$

Furthermore, the cross ratio is invariant under the action of $\mathrm{PGL}(\mathcal{R}, 2)$:

$$cr(A, B, C, D) = cr(A^\gamma, B^\gamma, C^\gamma, D^\gamma), \text{ for all } \gamma \in \mathrm{PGL}(\mathcal{R}, 2).$$

If $d \in \mathcal{C}(\mathcal{R})$ the conjugacy class $cr(A, B, C, D)$ becomes a single element and we write $cr(A, B, C, D) = d$. A theorem taken from [33] states:

Theorem 2.6. *(1) Four mutually distant points* $A, B, C, D \in \mathbb{P}^1(\mathcal{R})$ *are cocatenal, i.e., incident with one and the same chain, if* $cr(A, B, C, D) \in \mathcal{L}^\times$.

(2) Let A,B,C be mutually distant points. The chain containing A,B,C is the set

$$\{A\} \cup \{X \in \mathbb{P}^1(\mathcal{R}) \mid X \Delta A \text{ and } cr(A,B,C,X) \in \mathcal{L}\}.$$

Th. 2.6 allows a parametrisation of chains with the help of the cross ratio. Therefore, we take a closer look at the cross ratio. The map $\gamma \in \mathrm{PGL}(\mathcal{R},2)$ that takes three arbitrary mutually distant proper points $A,B,C \in \mathbb{P}^1(\mathcal{R})$ to the points U,V,W can be constructed in the following way:

$$\mathrm{M} = \mathrm{M}_1\mathrm{M}_2\mathrm{M}_3\mathrm{M}_4 = \begin{pmatrix} 1 & 0 \\ -c & 1 \end{pmatrix} \begin{pmatrix} 0 & 1 \\ 1 & 0 \end{pmatrix} \begin{pmatrix} 1 & 0 \\ t & 1 \end{pmatrix} \begin{pmatrix} v & 0 \\ 0 & 1 \end{pmatrix} = \begin{pmatrix} tv & 1 \\ (1-ct)v & -c \end{pmatrix},$$

with $t = (c-a)^{-1}$ and $v = (t+(b-c)^{-1})^{-1}$. In this product the first two matrices $\mathrm{M}_1\mathrm{M}_2$ put the point C to W, where it is fixed under the action of $\mathrm{M}_3\mathrm{M}_4$. The third factor M_3 maps the image of A under the first two factors to U. The last factor M_4 does not affect the image of A under the first three matrices. Finally, the last factor has traced B through the first three factors and maps the image of B to V. To check this we apply the mapping γ described by the matrix M to the three points:

$$\mathcal{R}(a,1)\mathrm{M} = \mathcal{R}(a,1) \begin{pmatrix} tv & 1 \\ (1-ct)v & -c \end{pmatrix} = \mathcal{R}(atv + (1-ct)v, a-c)$$
$$= \mathcal{R}((a-c)t+1)v, a-c) = \mathcal{R}((-1+1)v, a-c) \sim \mathcal{R}(0,1) = U,$$

$$\mathcal{R}(b,1)\mathrm{M} = \mathcal{R}(b,1) \begin{pmatrix} tv & 1 \\ (1-ct)v & -c \end{pmatrix} = \mathcal{R}(btv + (1-ct)v, b-c)$$
$$= \mathcal{R}((b-c)t+1)v, b-c) \sim \mathcal{R}((b-c)t+1, (b-c)v^{-1})$$
$$= \mathcal{R}((b-c)t+1, (b-c)t+1) \sim \mathcal{R}(1,1) = V,$$

$$\mathcal{R}(c,1)\mathrm{M} = \mathcal{R}(c,1) \begin{pmatrix} tv & 1 \\ (1-ct)v & -c \end{pmatrix} = \mathcal{R}(ctv + 1 - ctv, c-c)$$
$$= \mathcal{R}(1,0) = W.$$

Let us now consider the action on an arbitrary proper point $\mathcal{R}(x,1) \in \mathbb{P}^1(\mathcal{R})$:

2.2 Chain Geometry as Incidence Geometry

$$\mathcal{R}(x,1)\begin{pmatrix} tv & 1 \\ (1-ct)v & -c \end{pmatrix} = \mathcal{R}\left(((x-c)t+1)v, x-c\right)$$
$$= \mathcal{R}\left((x-c)(c-a)^{-1} + (c-a)(c-a)^{-1})v, x-c\right)$$
$$= \mathcal{R}\left(((x-c) + (c-a))(c-a)^{-1}v, x-c\right)$$
$$= \mathcal{R}\left((x-a)(c-a)^{-1}((c-a)^{-1} + (b-c)^{-1})^{-1}, x-c\right)$$
$$= \mathcal{R}\left(((c-a)(x-a)^{-1})^{-1}((c-a)^{-1} + (b-c)^{-1})^{-1}, x-c\right)$$
$$= \mathcal{R}\left((((c-a)^{-1} + (b-c)^{-1})((c-a)(x-a)^{-1}))^{-1}, x-c\right)$$
$$= \mathcal{R}\left(((x-a)^{-1} + (b-c)^{-1}(c-a)(x-a)^{-1})^{-1}, x-c\right)$$
$$= \mathcal{R}\left(((1 + (b-c)^{-1}(c-a))(x-a)^{-1})^{-1}, x-c\right)$$
$$= \mathcal{R}\left((((b-c)^{-1}(b-c) + (b-c)^{-1}(c-a))(x-a)^{-1})^{-1}, x-c\right)$$
$$= \mathcal{R}\left((((b-c)^{-1}((b-c) + (c-a))(x-a)^{-1})^{-1}, x-c\right)$$
$$= \mathcal{R}\left(((b-c)^{-1}(b-a)(x-a)^{-1})^{-1}, x-c\right)$$
$$= \mathcal{R}\left((x-a)(b-a)^{-1}(b-c), x-c\right)$$
$$= \mathcal{R}\bigl(\underbrace{(x-a)(b-a)^{-1}(b-c)}_{\mu}(1,0) + \underbrace{(x-c)}_{\lambda}(0,1)\bigr). \tag{2.1}$$

We repeat this calculation for an ideal point $\mathcal{R}(1,n)$, $n \in \mathcal{N}(\mathcal{R})$, where $\mathcal{N}(\mathcal{R})$ denotes the set zero divisors of $\mathcal{R}$.

$$\mathcal{R}(1,n)\begin{pmatrix} tv & 1 \\ (1-ct)v & -c \end{pmatrix} = \mathcal{R}\left(tv + n(1-ct)v, 1-nc\right)$$
$$= \mathcal{R}\left((t + n(1-ct))v, 1-nc\right)$$
$$= \mathcal{R}\left((t + n(t^{-1}t - ct))v, 1-nc\right)$$
$$= \mathcal{R}\left((1 + n(t^{-1} - c))tv, 1-nc\right)$$
$$= \mathcal{R}\left((1-na)tv, 1-nc\right)$$
$$= \mathcal{R}\left((1-na)(c-a)^{-1}((c-a)^{-1} + (b-c)^{-1})^{-1}, 1-nc\right)$$
$$= \mathcal{R}\left(((c-a)(1-na)^{-1})^{-1}((c-a)^{-1} + (b-c)^{-1})^{-1}, 1-nc\right)$$
$$= \mathcal{R}\left((((c-a)^{-1} + (b-c)^{-1})(c-a)(1-na)^{-1})^{-1}, 1-nc\right)$$
$$= \mathcal{R}\left(((1 + (b-c)^{-1}(c-a))(1-na)^{-1})^{-1}, 1-nc\right)$$
$$= \mathcal{R}\left((((b-c)^{-1}(b-c)+(b-c)^{-1}(c-a))(1-na)^{-1})^{-1}, 1-nc\right)$$
$$= \mathcal{R}\left(((b-c)^{-1}(b-a)(1-na)^{-1})^{-1}, 1-nc\right)$$
$$= \mathcal{R}\left((1-na)(b-a)^{-1}(b-c), 1-nc\right)$$

$$= \mathcal{R}\big(\underbrace{(1-na)(b-a)^{-1}(b-c)}_{\mu}(1,0) + \underbrace{(1-nc)}_{\lambda}(0,1)\big). \tag{2.2}$$

With the upper calculation we are able to compute the cross ratio as

$$f(X) = cr(A,B,C,X) = \mathcal{R}(x_1,x_0)M = \mathcal{R}(\mu,\lambda).$$

If the cross ratio is a proper element of the projective line, we can find normalized representatives of the equivalence class:

X proper: $\mathcal{R}(\mu,\lambda) \sim \mathcal{R}((x-a)(b-a)^{-1}(b-c)(x-c)^{-1}, 1),$ (2.3)
X ideal: $\mathcal{R}(\mu,\lambda) \sim \mathcal{R}((1-na)(b-a)^{-1}(b-c)(1-nc)^{-1}, 1).$ (2.4)

If the cross ratio is an ideal element we can normalize to

X proper: $\mathcal{R}(\mu,\lambda) \sim \mathcal{R}(1, (x-c)(b-c)^{-1}(b-a)(x-a)^{-1}),$ (2.5)
X ideal: $\mathcal{R}(\mu,\lambda) \sim \mathcal{R}(1, (1-nc)(b-c)^{-1}(b-a)(1-na)^{-1}).$ (2.6)

Remark 2.9. *For a proper cross ratio $\mathcal{R}(\Lambda,1)$ we can calculate the cross ratio directly from the ring elements by*

$$cr(a,b,c,x) = (x-a)(b-a)^{-1}(b-c)(x-c)^{-1}. \tag{2.7}$$

This is exactly the cross ratio formula that is used for complex numbers or in classical projective geometry. If $\Lambda \in C(\mathcal{R})$ the cross ratio is an element and not a conjugacy class.

With the help of Th. 2.6 and the formulas (2.3)-(2.6) we can now parametrise chains in the chain geometry $\Sigma(\mathcal{L},\mathcal{R})$

$$(d-a)(b-a)^{-1}(b-c)(d-c)^{-1} = \Lambda$$
$$(d-a)(b-a)^{-1}(b-c) = \Lambda(d-c)$$
$$d(b-a)^{-1}(b-c) - a(b-a)^{-1}(b-c) = \Lambda d - \Lambda c$$
$$d(b-a)^{-1}(b-c) - \Lambda d = a(b-a)^{-1}(b-c) - \Lambda c$$
$$d\left[(b-a)^{-1}(b-c) - \Lambda\right] = a(b-a)^{-1}(b-c) - \Lambda c$$
$$d = \left[a(b-a)^{-1}(b-c) - \Lambda c\right]\left[(b-a)^{-1}(b-c) - \Lambda\right]^{-1}. \tag{2.8}$$

2.2 Chain Geometry as Incidence Geometry

Properties of the Cross Ratio Let $a, b, c \in \mathcal{R}$ be three pairwise distant elements and let $d \in \mathcal{R}$ with $d \Delta c$. Furthermore, we assume that the cross ratio is proper and not a conjugacy class. Therefore, we can use Eq. (2.7) to compute the cross ratio. It follows

$$cr(a, b, c, a) = (a-a)(b-a)^{-1}(b-c)(a-c)^{-1} = 0,$$
$$cr(a, b, c, b) = (b-a)(b-a)^{-1}(b-c)(b-c)^{-1} = 1.$$

and with Eq. (2.1)

$$cr(a, b, c, c) = \big((c-a)(b-a)^{-1}(b-c)(1,0) + (c-c)(0,1)\big)$$
$$= (1, 0) = \infty.$$

Thus, we can use chains parametrised by the cross ratio for interpolation purposes, since the elements a, b, c are interpolated for the cross ratio values $0, 1, \infty$.

Theorem 2.7. *Let $a, b, c, d \in \mathcal{R}$ be four mutually distant elements incident with one chain. The exchange of two pairs of elements does not change the value of the cross ratio:*

$$cr(a, b, c, d) = cr(b, a, d, c) = cr(c, d, a, b) = cr(d, c, b, a).$$

All in all there are 24 permutations of the elements a, b, c, d. If we use the symmetries given above, we have six possible values for the cross ratios:

$$\begin{aligned}
cr(a, b, c, d) &= \Lambda, & cr(a, d, c, b) &= \Lambda^{-1} \\
cr(a, c, d, b) &= (1-\Lambda)^{-1}, & cr(b, a, c, d) &= 1-\Lambda \\
cr(a, c, b, d) &= \Lambda(\Lambda-1)^{-1} & cr(a, d, b, c) &= (\Lambda-1)\Lambda^{-1}.
\end{aligned} \qquad (2.9)$$

Note, that $\Lambda \in \mathcal{C}(\mathcal{R})$ if a, b, c, d are cocatenal.

Proof. The proof is done be direct computation. We restrict ourselves to the case where four elements are units. Then, Eq. (2.1) is valid. If there are zero divisors, we have to take Eq. (2.4). The first four symmetries can be proofed in the same way. As an example we show $cr(a, b, c, d) = cr(c, d, a, b)$ and we use that $cr(a, b, c, d) \in \mathcal{C}(\mathcal{R})$:

$$\begin{aligned}
cr(a,b,c,d) &= (d-a)(b-a)^{-1}(b-c)(d-c)^{-1} \\
&= (d-a)(b-a)^{-1}(b-c)(d-c)^{-1} \cdot 1 \\
&= (d-a)(b-a)^{-1}(b-c)(d-c)^{-1} \\
&\quad (b-c)(d-c)^{-1}(d-c)(b-c)^{-1} \\
&= (b-c)(d-c)^{-1}(d-a)(b-a)^{-1} \\
&\quad (b-c)(d-c)^{-1}(d-c)(b-c)^{-1} \\
&= (b-c)(d-c)^{-1}(d-a)(b-a)^{-1} \\
&= cr(c,d,a,b).
\end{aligned}$$

Now we show $cr(a,b,c,d)^{-1} = cr(c,b,a,d)$ which is also done by direct computation.

$$\begin{aligned}
cr(a,b,c,d)^{-1} &= \left[(d-a)(b-a)^{-1}(b-c)(d-c)^{-1}\right]^{-1} \\
&= \left((b-c)(d-c)^{-1}\right)^{-1}\left((d-a)(b-a)^{-1}\right)^{-1} \\
&= (d-c)(b-c)^{-1}(b-a)(d-a)^{-1} \\
&= cr(c,b,a,d)
\end{aligned}$$

To complete the proof it is sufficient to show one more of the identities (2.9), since all other identities can be achieved by combining two identities. This can be done be direct calculation and is left to reader as an exercise. □

2.3 Quadric Chain Spaces

Before we define chain spaces on quadrics we give a short introduction to quadrics in general, cf. [25].

Definition 2.15. *Let V be a vector space over a commutative field $\mathcal{K}$. A map $\rho: V \to \mathcal{K}$ is called a quadratic form, if for all $x,y \in V$, $\lambda \in \mathcal{K}\setminus\{0\}$:*

(1) $\rho(\lambda x) = \lambda^2 \rho(x)$.

(2) $(x,y) \mapsto \rho(x+y) - \rho(x) - \rho(y)$ *is a bilinear form* $\sigma: V \times V \to \mathcal{K}$.

2.3 Quadric Chain Spaces

Remark 2.10. *For a bilinear form* $\sigma : V \times V \to \mathcal{K}$ *we can explain* ρ *as*

$$\sigma(x, x) =: \rho(x).$$

Furthermore, we can define the bilinear form with

$$\rho(x + y) = \sigma(x + y, x + y) = \sigma(x, x) + \sigma(x, y) + \sigma(y, x) + \sigma(y, y)$$
$$= \rho(x) + 2\sigma(x, y) + \rho(y).$$

If char $\mathcal{K} \neq 2$, *we can define the symmetric bilinear form*

$$\sigma_\rho : V \times V \to \mathcal{K} : (x, y) \mapsto \frac{1}{2}(\rho(x + y) - \rho(x) - \rho(y)).$$

In the following we assume char $\mathcal{K} \neq 2$.

Each quadratic form ρ can be written as $\rho(x) = x^T A x$, with $A = A^T$. The corresponding bilinear form reads $\sigma(x, y) = x^T A y$.

Definition 2.16. *Let* $\mathbb{P}^n(V)$ *be an n-dimensional projective space over* $\mathcal{K}$ *and let* $\rho : V \to \mathcal{K}$ *be a quadratic form.*

(1) The set $\mathfrak{Q} := \mathfrak{Q}(\rho) = \{x\mathcal{K} \in \mathbb{P}^n(\mathcal{K}) \mid \rho(x) = 0\}$ *is called the projective quadric corresponding to* ρ. *This is well-defined since* $\rho(x) = 0$ *implies that* $\rho(xk) = 0$, $k \in \mathcal{K}$.

(2) Let l be a line in $\mathbb{P}^n(\mathcal{K})$. *If* $|l \cap \mathfrak{Q}| = 1$, l *is called* tangent. *If* $|l \cap \mathfrak{Q}| = 2$, l *is called* secant. *If the intersection consists of the whole line, the line is tangent to and contained in the quadric.*

(3) Let p be a point contained by $\mathfrak{Q}$. The set

$$\mathfrak{T}_p := \{q \in l \mid l \text{ tangent to } \mathfrak{Q} \text{ at the point } p\}$$

is called tangent space *to* $\mathfrak{Q}$ *at the point p.*

(4) The symmetric matrix A corresponding to the quadratic form ρ defines a correlation in $\mathbb{P}^n(\mathcal{K})$, *called a* polarity. *All points that are incident with their image hyperplane are contained in* $\mathfrak{Q}$.

Remark 2.11. *With the polarity given by* A, *a point* $p \in \mathfrak{Q}$ *is mapped to the tangent hyperplane at the point p. Therefore, we can use the polarity to describe tangent spaces. If the tangent space at a point $s \in \mathfrak{Q}$ is the whole space, s is called* double point *of $\mathfrak{Q}$. $\mathfrak{T}_p$ is a hyperplane if the point $p \in \mathfrak{Q}$ is not a double point. A line through a given point $p \in \mathfrak{Q}$ is either a tangent or a secant.*

With these preliminaries we want to define a chain space on a quadric. We follow [11] and assume $\dim V > 4$. Furthermore, we demand that the quadric $\mathfrak{Q}$ has at least one secant and that it is not the union of two hyperplanes.

Definition 2.17. *Let $\mathfrak{Q}^*$ be the set of points of $\mathfrak{Q}$ that are not double points. This set becomes the point set $\mathcal{P}$ of the chain geometry. A plane E is called* admissible *if the intersection $E \cap \mathfrak{Q}$ contains at least three points, but no whole line of $\mathfrak{Q}$. The set of all conic sections with admissible planes is denoted by*

$$\mathfrak{C}(\mathfrak{Q}) := \{\mathfrak{Q} \cap E \mid E \text{ admissible}\}$$

and becomes the chain set. Hence, we examine the incidence structure

$$\Sigma(\mathfrak{Q}) := (\mathfrak{Q}^*, \mathfrak{C}(\mathfrak{Q})).$$

Theorem 2.8. *The incidence structure $\Sigma(\mathfrak{Q}) := (\mathfrak{Q}^*, \mathfrak{C}(\mathfrak{Q}))$ is a chain space.*

To proof this theorem, we refer to [11]. We just have to show the axioms **C1-C4**.

2.4 Real Benz Planes

In this section we give famous examples for chain geometries. Hence, we examine chain geometries over two-dimensional algebras over the real numbers. The famous textbook [5] treated this topic exhaustively. Furthermore, we present the related quadric models. This section follows the Wikipedia article about Benz planes [67].

There are only three different cases of two-dimensional algebras over $\mathbb{R}$. The algebras are the complex numbers $\mathbb{C}$, the dual numbers $\mathbb{D}$, and

2.4 Real Benz Planes

the double numbers also called the split-complex numbers $\mathbb{A} = \mathbb{R} \times \mathbb{R}$. Thus, we have to examine the three different chain geometries $\Sigma(\mathbb{R},\mathbb{C})$, $\Sigma(\mathbb{R},\mathbb{D})$, and $\Sigma(\mathbb{R},\mathbb{A})$.

Theorem 2.9. *The chain geometry $\Sigma(\mathbb{R},\mathbb{K})$ is a*

(1) Möbius geometry, if $\mathbb{K} = \mathbb{C}$,

(2) Laguerre geometry, if $\mathbb{K} = \mathbb{D}$,

(3) Minkowski geometry, if $\mathbb{K} = \mathbb{A}$.

A proof can be found in [11]. All these planes arise from the real affine plane with different compactifications. The Möbius plane is the affine plane with an additional point called ∞, *i.e.*, the conformally closed affine plane. Chains are circles and lines that can be interpreted as circles with infinite radius. A corresponding quadric model is the unit sphere with its planar sections as set of chains. The stereographic projection provides a mapping between the planar model and the quadric model. Chains in the Laguerre plane are parabolas of the form $y = ax^2+bx+c$ or lines, if $a = 0$. Here the real affine plane is closed with a line at infinity, *i.e.*, the projective closure. To every parabola $y = ax^2+bx+c$ we have to add the point (∞, a), and therefore, the point $\infty := (\infty, \infty)$ is also contained on the ideal line. As adequate quadric model we have the so called Blaschke cylinder with its planar sections as chains and as a mapping between the planar and the quadric model a generalized stereographic projection from the plane to the cylinder. In the Minkowski plane $\Sigma(\mathbb{R}, \mathbb{A})$ the chains are hyperbolas of the form $y = \frac{a}{x-b}+c$, $a \neq 0$ and lines $y = mx + n$, $m \neq 0$. To every line we add the point (∞, ∞) and to every hyperbola we add the two points (b, ∞), (∞, c). Via a generalized stereographic projection we can map the planar sections of a one-sheeted hyperboloid to this set of chains.

Remark 2.12. *All these chains can be parametrised with Eq. (2.8) and with three algebra elements corresponding to three points the circle, the parabola, the hyperbola, or the line is determined uniquely.*

2.5 Jordan-Systems

Now we take a look at sub structures of chain geometries.

Definition 2.18. *Let $\Sigma = (\mathcal{P}, \mathfrak{C})$ be a chain space and let $\mathcal{S} \subseteq \mathcal{P}$ with*

(1) $\mathcal{S}$ contains three mutually distant points.

(2) If $p, q, r \in \mathcal{S}$ are mutually distant, then the connecting chain $c = (pqr)$ is entirely contained in $\mathcal{S}$. The set of all connecting chains is denoted by $\mathfrak{C}(\mathcal{S})$.

(3) The incidence structure $(\mathcal{S}, \mathfrak{C}(\mathcal{S}))$ is a chain space.

Under this conditions $(\mathcal{S}, \mathfrak{C}(\mathcal{S}))$ is a subspace *of Σ. If $(\mathcal{S}, \mathfrak{C}(\mathcal{S}))$ satisfies just condition (2), $(\mathcal{S}, \mathfrak{C}(\mathcal{S}))$ is called a* weak subspace *of Σ.*

Trivial subspaces of $\Sigma = \Sigma(\mathcal{P}, \mathfrak{C})$ are Σ itself and subspaces containing just a single chain $c \in \mathfrak{C}$. More examples can be generated with subalgebras.

Definition 2.19. *Let $\mathcal{R}$ be an algebra over the field $\mathcal{K}$. A subspace $\mathcal{S}$ of $\mathcal{R}$ is called a* subalgebra *of $\mathcal{R}$, if*

(1) $1 \in \mathcal{S}$.

(2) For $a, b \in \mathcal{S}$ the product ab is in $\mathcal{S}$.

(3) If $a \in \mathcal{S} \cap \mathcal{R}^\times$, then $a^{-1} \in \mathcal{S}$, i.e, $\mathcal{S} \cap \mathcal{R}^\times = \mathcal{S}^\times$.

Since chain geometries over algebras are chain spaces, the chain geometries over subalgebras are also chain spaces, and therefore, subspaces of the chain space. It is not common to demand condition (3). For the description of subspaces of chain spaces this condition is important. Just with this condition subalgebras correspond to Jordan-systems. The following theorem states that subalgebras also define chain geometries on their own.

Theorem 2.10. *Let $\mathcal{R}$ be a $\mathcal{K}$-algebra and let $\mathcal{S}$ be a subalgebra of $\mathcal{R}$. The mapping $\iota: \mathbb{P}^1(\mathcal{S}) \to \mathbb{P}^1(\mathcal{R})$ with $\mathcal{S}(a, b) \mapsto \mathcal{R}(a, b)$ is an injective morphism of chain spaces $\Sigma(\mathcal{K}, \mathcal{S}) \to \Sigma(\mathcal{K}, \mathcal{R})$. The image $\iota(\Sigma(\mathcal{K}, \mathcal{S}))$ is a subspace of $\Sigma(\mathcal{K}, \mathcal{R})$, that is embedded via ι and isomorphic to $\Sigma(\mathcal{K}, \mathcal{S})$.*

2.5 Jordan-Systems

For a proof we refer to [11].

Remark 2.13. *We identify $\iota(\Sigma(\mathcal{K}, \mathcal{S}))$ with $\Sigma(\mathcal{K}, \mathcal{S})$.*

Example 2.1. *As example we take the skew field of quaternions, see section 1.1.2. A general quaternion has the form*

$$q = a + b\mathbf{i} + c\mathbf{j} + d\mathbf{k}, \quad a, b, c, d \in \mathbb{R}.$$

A subalgebra is defined by the set $\{a + c\mathbf{j} \mid a, c \in \mathbb{R}\}$. In fact this subalgebra is isomorphic to the complex numbers $\mathbb{C}$ and it defines a subspace $\Sigma(\mathbb{R}, \mathbb{C})$ of the chain geometry $\Sigma(\mathbb{R}, \mathbb{H})$.

There are also subspaces of chain spaces that do not correspond to subalgebras. To describe some of these subspaces we need the definition of a Jordan-system.

Definition 2.20. *Let $\mathcal{R}$ be a $\mathcal{K}$-algebra and let $\mathcal{J}$ be a subspace of the $\mathcal{K}$-vector space $\mathcal{R}$ with $1 \in \mathcal{J}$. The subspace $\mathcal{J}$ is called*

J1 *Jordan-system in $\mathcal{R}$, if $b \in \mathcal{J} \cap \mathcal{R}^\times \Rightarrow b^{-1} \in \mathcal{J}$ for all $b \in \mathcal{J}$, we define $\mathcal{J}^\times = \mathcal{J} \cap \mathcal{R}^\times$.*

J2 *Jordan-closed in $\mathcal{R}$, if $a, b \in \mathcal{J} \Rightarrow aba \in \mathcal{J}$ for all $a, b \in \mathcal{J}$.*

J3 *strong in $\mathcal{R}$, if for all $b \in \mathcal{J}$ the condition $|e(b)| > |\mathcal{K} \backslash e(b)|$ is satisfied, where*

$$e(x) := \{f \in \mathcal{K} \mid x + f \in \mathcal{R}^\times\}.$$

Remark 2.14. *If $\mathcal{R}$ is a strong algebra, cf. Def. 2.4, then every Jordan-system is strong in $\mathcal{R}$.*

Lemma 2.2. *Let $\mathcal{J}$ be a Jordan-closed Jordan-system. For $a, b \in \mathcal{J}$ the elements a^2 and $ab + ba$ are also elements of $\mathcal{J}$. The set $\mathcal{J}$ together with $a \circ b = \frac{1}{2}(ab + ba)$ is called* Jordan algebra *if char $\mathcal{K} \neq 2$.*

Remark 2.15. *The term Jordan-algebra is named after the German physicist P.* JORDAN.

Lemma 2.3. *Let $\mathcal{J}$ be a strong Jordan-system in $\mathcal{R}$. Then $\mathcal{J}$ is also Jordan-closed in $\mathcal{R}$.*

Now we define the projective line over the Jordan-system $\mathcal{J}$.

Definition 2.21. *Let $\mathcal{J}$ be a strong Jordan-system in the $\mathcal{K}$-algebra $\mathcal{R}$. We define the* projective line over $\mathcal{J}$ *as*

$$\mathbb{P}^1(\mathcal{J}) := \{\mathcal{R}(1+ab, a) \mid a, b \in \mathcal{J}\}.$$

There is a subgroup of $\mathrm{PGL}(\mathcal{R}, 2)$ that leaves $\mathbb{P}^1(\mathcal{J})$ invariant.

Definition 2.22. *Let $\Delta(\mathcal{J})$ be the subgroup of $\mathrm{PGL}(\mathcal{R}, 2)$ that is generated by the matrices $\begin{pmatrix} 1 & 0 \\ q & 1 \end{pmatrix}, q \in \mathcal{J}$ and $\begin{pmatrix} 0 & 1 \\ 1 & 0 \end{pmatrix}$. Then $\Delta(\mathcal{J})$ is a subgroup of $\mathrm{Aut}(\mathbb{P}^1(\mathcal{R}), \Delta)$ that leaves $\mathbb{P}^1(\mathcal{J})$ invariant.*

Theorem 2.11. *Let $\mathcal{J}$ be a strong Jordan-system in $\mathcal{R}$. If there are three mutually distant points $p, q, r \in \mathbb{P}^1(\mathcal{J})$, their common chain $c = (pqr) \in \mathfrak{C}(\mathcal{K}, \mathcal{R})$ lies completely in $\mathbb{P}^1(\mathcal{J})$. Every such chain can be written as $c = c_h^\delta$, where $\delta \in \Delta(\mathcal{J}), h \in \mathcal{J}^\times$ and $c_h := \{\mathcal{R}(sh, t) \mid \mathcal{K}(s, t) \in \mathbb{P}^1(\mathcal{K})\}$. Conversely, for $\delta \in \Delta(\mathcal{J}), h \in \mathcal{J}^\times$ the chain c_h^δ is contained in $\mathbb{P}^1(\mathcal{J})$.*

Proof. For a proof we refer to [11]. □

It can be shown that the subspace $\mathbb{P}^1(\mathcal{J})$ is a subspace of $\Sigma(\mathcal{K}, \mathcal{R})$, see [11]. As distance relation on $\mathbb{P}^1(\mathcal{J})$ we use the distance relation Δ of $\mathbb{P}^1(\mathcal{R})$.

Theorem 2.12. *Let $\mathcal{J}$ be a strong Jordan-system in $\mathcal{R}$. Then*

$$\Sigma' = \Sigma(\mathcal{K}, \mathcal{R}, \mathcal{J}) = \Sigma(\mathbb{P}^1(\mathcal{J}), \mathfrak{C}(\mathcal{K}, \mathcal{R}, \mathcal{J}))$$

with $\mathfrak{C}(\mathcal{K}, \mathcal{R}, \mathcal{J}) = \{c_h^\delta \mid h \in \mathcal{J}^\times, \delta \in \Delta(\mathcal{J})\}$ is a subspace of $\Sigma = \Sigma(\mathcal{K}, \mathcal{R})$. We call $\Sigma(\mathcal{K}, \mathcal{R}, \mathcal{J})$ the chain geometry over $(\mathcal{K}, \mathcal{R}, \mathcal{J})$.

Now we generalize the setting and assume that the set $\mathcal{S}$ does not contain the three points $\mathcal{R}(1, 0)$, $\mathcal{R}(0, 1)$, and $\mathcal{R}(1, 1)$. Under these assumptions $\mathcal{S}$ cannot be the projective line over a Jordan-system, see Def. 2.20. Nevertheless, we can formulate the following theorem, see [11]:

Theorem 2.13. *Let $\mathcal{R}$ denote a strong $\mathcal{K}$-algebra, $\Sigma = \Sigma(\mathcal{K}, \mathcal{R})$ and $\mathcal{S}$ a non-trivially connected weak subspace of Σ. Under this conditions there exists $\gamma \in \mathrm{PGL}(\mathcal{R}, 2)$ with $\mathcal{S}^\gamma = \mathbb{P}^1(\mathcal{J})$ for a strong Jordan-system $\mathcal{J}$ of $\mathcal{R}$. The subspace $(\mathcal{S}, \mathfrak{C}(\mathcal{S}))$ is a strong subspace that is isomorphic to $\Sigma(\mathcal{K}, \mathcal{R}, \mathcal{J})$.*

Proof. The theorem follows directly from the 3-Δ-transitivity of $\mathrm{PGL}(\mathcal{R}, 2)$, cf. [11]. □

2.6 Contact Spaces

Contact spaces are of certain interest on their own. BENZ introduced contact spaces and a contact relation for two-dimensional commutative algebras, see [5]. Furthermore, a survey-like description can be found in [33]. The contact relation gives a method to compute chains that are in contact with other chains at a certain point by means of the cross ratio. First of all, we define the contact relation following [33].

Definition 2.23. *Let $\Sigma = (\mathcal{P}, \mathfrak{C})$ be a chain geometry. A contact relation on Σ is a family (ρ_p), $p \in \mathcal{P}$, where each (ρ_p) is an equivalence relation on $(p) := \{c \in \mathfrak{C} \mid p \in c\}$ with the following conditions (we write aPb for $a\rho_p b$):*

(1) aPb and $a \neq b$ implies $|a \cap b| = 1$,

(2) for $a \in (p)$ and $q \in \mathcal{P}$ with $p \not\Vert q$ there is exactly one $b \in (p)$ with $q \in b$ and aPb.

A chain geometry, i.e., a chain space together with a contact relation is called a contact space.

Remark 2.16. *This definition is also adequate for weak chain spaces.*

On the partial affine space Σ_p the contact relation ρ_p induces a parallelism. In every chain space $\Sigma = (\mathcal{P}, \mathfrak{C})$ we have the following natural contact relation:

$$\forall p \in \mathcal{P} : aPb \Leftrightarrow a\setminus\{p\} \text{ and } b\setminus\{p\} \text{ are parallel lines in } \Sigma_p.$$

BENZ [5] defined a contact relation with the help of the residual spaces without mentioning them explicitly. We also give his definition since it is more intuitive.

Definition 2.24. *Let $\Sigma = (\mathcal{P}, \mathfrak{C})$ be a chain geometry, let $W = \mathcal{R}(1,0) \in \mathbb{P}^1(\mathcal{R})$ and let $a, b \in \mathfrak{C}$. The chains a and b are in contact at the point P if there is a $\gamma \in \mathrm{PGL}(\mathcal{R}, 2)$, such that $P^\gamma = W$ and the lines $a^\gamma \backslash \{W\}$, $b^\gamma \backslash \{W\}$ are parallel.*

We give a theorem from [5]:

Theorem 2.14. *Let aPb, $a \neq b$. It follows $a \cap b = P$. The contact relation ρ_p is an equivalence relation, i.e., it is reflexive, symmetric, and transitive.*
Contact theorem: *Let $c \in \mathfrak{C}$ be a chain and let $p \in c$, $q \notin c$ be points with $p \not\parallel q$. Then there is exactly one chain c' incident with p and q that is in contact with c.*

Lemma 2.4. *If aPb and $\gamma \in \mathrm{PGL}(\mathcal{R}, 2)$, then $a^\gamma P^\gamma b^\gamma$.*

The inverse implication is not true in general, but, if there is a so-called beetle-figur (german: Käferfigur), then it is true. If $\Sigma = (\mathcal{P}, \mathfrak{C})$ is a chain space, then there is necessarily a beetle-figure. Now we give two theorems that connect the contact relation with the cross ratio. Both theorems can be found in [5].

Theorem 2.15. *Let $\mathcal{R}$ be a $\mathcal{K}$-algebra over a field and let $a, b \in \mathfrak{C}$ be chains incident with $P \in \mathbb{P}^1(\mathcal{R})$. The following expressions are equivalent:*

(1) aPb,

(2) For two different points $A, A' \in a \backslash \{P\}$ and for two points $B, B' \in b \backslash \{P\}$
$$cr(P, A, A', B) - cr(P, A, A', B') \in \mathcal{K},$$

(3) There are two different points $A, A' \in a \backslash \{P\}$ and two different points $B, B' \in b \backslash \{P\}$ with
$$cr(P, A, A', B) - cr(P, A, A', B') \in \mathcal{K}.$$

The next theorem allows a parametrisation of the chain b that is in contact to another chain a at the point P.

Theorem 2.16. *Let $P \in a$, $Q \notin a$, $P \not\parallel Q$, then there exists exactly one chain b incident with P and Q that is in contact with a. This chain consists of all points $X \not\parallel P$ with*

$$cr(P, A, A', X) - cr(P, A, A', Q) \in \mathcal{K},$$

where A, A' are different points on $a \setminus \{P\}$. Therefore,

$$b = \{P\} \cup \{X \not\parallel P \mid cr(P, A, A', X) - cr(P, A, A', Q) \in \mathcal{K}\}.$$

A proof can be found in [5]. The proof given there is valid for commutative rings, but every statement can be transferred directly to the non-commutative case. The proof uses Lemma 2.4 and the invariance of the cross ration under the action of the group $\mathrm{PGL}(\mathcal{R}, 2)$. With this theorem we are able to calculate an explicit parametrisation of the chain that is in contact to a given one.

2.7 Chain Geometries over Clifford Algebras

Benz planes can be interpreted as chain geometries over two-dimensional Clifford algebras, i.e., chain geometries over $\mathcal{C}\ell_{(1,0,0)} \cong \mathbb{C}$, $\mathcal{C}\ell_{(0,1,0)} \cong \mathbb{A}$, or $\mathcal{C}\ell_{(0,0,1)} \cong \mathbb{D}$. Furthermore, we are dealing with finite-dimensional Clifford algebras over the real numbers. For these algebras the chain geometry is well understood, and therefore, we can apply the methods easily, see [5]. Every Clifford algebra is an algebra over a commutative ring. The dimension of the center depends on the dimension of the algebra, see Def. 1.11. The real numbers are always contained in the center. Therefore, we can always define a chain geometry $\Sigma(\mathbb{R}, \mathcal{C}\ell_{(p,q,r)})$. If the center of the Clifford algebra is two-dimensional, it is isomorphic to $\mathbb{A}, \mathbb{C}$, or $\mathbb{D}$. In these cases we can define different chain geometries. On one hand, we have $\Sigma(\mathbb{R}, \mathcal{C}\ell_{(p,q,r)})$ and on the other hand $\Sigma(\mathcal{R}, \mathcal{C}\ell_{(p,q,r)})$ where $\mathcal{R}$ is $\mathbb{A}, \mathbb{C}$, or $\mathbb{D}$. The set of points of the chain geometry is the projective line over the Clifford algebra

$$\mathbb{P}^1(\mathcal{R}) := \left\{ \mathcal{R}(a,b) \mid \exists c, d : \begin{pmatrix} a & b \\ c & d \end{pmatrix} \in \mathrm{GL}(\mathcal{R}, 2) \right\},$$

where $\mathcal{R}$ denotes the Clifford algebra. The chains are defined as $\mathbb{R}$-chains if not stated otherwise:

$$\mathfrak{C}(\mathbb{R},\mathcal{R}) := \left\{ \left(\mathbb{P}^1(\mathbb{R})\right)^\gamma \mid \gamma \in \mathrm{PGL}(\mathcal{R},2) \right\}.$$

Remark 2.17. *Especially for dual unit quaternions, chains over the center $C(\mathbb{H}_d) \cong \mathbb{D}$ are of interest, because the cross ratio of four elements of one chain results in a dual number.*

2.7.1 Grade-1 Subspace

If we are dealing with a not fully degenerate Clifford algebra, we can show that the $\bigwedge^1 V$ subspace defines a chain geometry. Therefore, we proof that another subspace of the Clifford algebra defines a Jordan-system. After that we use Lemma 2.1 to map the Jordan-system to the $\bigwedge^1 V$ subspace. Hence, we proof the following:

Theorem 2.17. *The subspace $\mathcal{I}$ of $\mathcal{Cl}_{(p,q,r)}$ generated by the set $\{e_0, e_{12}, e_{13}, \ldots e_{1n}\}$ is a Jordan-closed and strong Jordan-system.*

Proof. To show this we have to check **J1-J3**.

J1 Since every element of $\mathcal{I}$ can be written as the product of two vectors, it is a versor. The inverse element for a versor $\mathfrak{v}$ is given by $\mathfrak{v}^{-1} = \frac{v^*}{vv^*}$.

J2 For $n=1$ the theorem is clear. Our **base case** is $n=2$. Let $q = \begin{pmatrix} q_1 & 0 \\ 0 & q_2 \end{pmatrix}$ be the quadratic form that defines the Clifford algebra with $q_1, q_2 \in \{-1, 1, 0\}$. The set generated by $\mathcal{I} = \{e_0, e_{12}\}$ is the set that shall be a Jordan-system. Two elements $\mathfrak{a} = x_1 e_0 + x_2 e_{12}$ and $\mathfrak{b} = y_1 e_0 + y_1 e_{12}$ satisfy **J2**:

$$\mathfrak{aba} = (x_1^2 y_1 - (2x_1 x_2 y_2 + x_2^2 y_1)q_1 q_2)e_0 + (2x_1 x_2 y_1 + x_1^2 y_2 - x_2^2 y_2 q_1 q_2)e_{12}.$$

Inductive step: Let $q = \mathrm{diag}(q_1, q_2, \ldots, q_n)$ be the diagonal matrix defining the qua-dratic form with $q_1, \ldots, q_n \in \{-1, 1, 0\}$. Let

$$\mathfrak{a} = \underbrace{x_1 e_0 + x_2 e_{12} + \ldots + x_{n-1} e_{1(n-1)}}_{\mathfrak{A}} + x_n e_{1n},$$

2.7 Chain Geometries over Clifford Algebras

$$\mathfrak{b} = \underbrace{y_1 e_0 + y_2 e_{12} + \ldots + y_{n-1} e_{1(n-1)}}_{\mathfrak{B}} + y_n e_{1n}.$$

Now we verify that the Jordan-system is Jordan-closed:

$$\mathfrak{a}\mathfrak{b}\mathfrak{a} = (\mathfrak{A} + x_n e_{1n})(\mathfrak{B} + y_n e_{1n})(\mathfrak{A} + x_n e_{1n}) \tag{2.10}$$
$$= \mathfrak{A}\mathfrak{B}\mathfrak{A} + \mathfrak{A} y_n e_{1n} \mathfrak{A} + x_n e_{1n} \mathfrak{B} \mathfrak{A} + x_n y_n e_{1n} e_{1n} \mathfrak{A} + \mathfrak{A}\mathfrak{B} x_n e_{1n}$$
$$+ \mathfrak{A} y_n e_{1n} x_n e_{1n} + x_n e_{1n} \mathfrak{B} x_n e_{1n} + x_n y_n e_{1n} e_{1n} x_n e_{1n}.$$

The statement is true for $n-1$, and therefore, $\mathfrak{A}\mathfrak{B}\mathfrak{A} \in \mathcal{I}$. It remains to show that the other terms are also in $\mathcal{I}$. We proof by direct computation that all terms

$$x_n y_n e_{1n} e_{1n} x_n e_{1n} = -x_n^2 y_n q_1 q_n e_{1n} \in \mathcal{I},$$
$$x_n e_{1n} \mathfrak{B} x_n e_{1n} = (-x_n^2 y_1 e_0 + x_n^2 y_2 e_{12} + \ldots + x_n^2 y_{n-1} e_{1(n-1)}) q_1 q_n$$
$$\mathfrak{A} y_n e_{1n} x_n e_{1n} = (-x_1 e_0 - x_2 e_{12} - \ldots - x_{n-1} e_{1(n-1)}) x_n y_n q_1 q_n,$$
$$x_n y_n e_{1n} e_{1n} \mathfrak{A} = (-x_1 e_0 - x_2 e_{12} - \ldots - x_{n-1} e_{1(n-1)}) x_n y_n q_1 q_n,$$
$$\mathfrak{A} y_n e_{1n} \mathfrak{A} = (x_1^2 + x_2^2 q_1 q_2 \ldots + x_{n-1}^2 q_1 q_{n-1}) y_n e_{1n},$$
$$x_n e_{1n} \mathfrak{B}\mathfrak{A} + \mathfrak{A}\mathfrak{B} x_n e_{1n} = 2(x_1 y_1 - x_2 y_2 q_1 q_2 - \ldots - x_{n-1} y_{n-1} q_1 q_{n-1}) x_n e_{1n}$$

are contained in $\mathcal{I}$. All in all we get $\mathfrak{a}\mathfrak{b}\mathfrak{a} \in \mathcal{I}$. Hence, $\mathcal{I}$ is Jordan-closed.

J3 We show that $|e(\mathfrak{a})| > |\mathbb{R}\backslash e(\mathfrak{a})|$. All elements $\mathfrak{a} - r$ with $\mathfrak{a} \in \mathcal{I}$ and $r \in \mathbb{R}$ that are zero divisors can be found by

$$(\mathfrak{a} - r)(\mathfrak{a} - r)^* = 0.$$

This defines a quadratic equation in r and has two solutions at the most. Hence, $|\mathbb{R}\backslash e(\mathfrak{a})| = 2$ and condition **J3** is satisfied.

$\square$

Remark 2.18. *The group* $\mathrm{PGL}(\mathcal{R}, 2)$ *acts flag-transitive on the set of flags* $\mathfrak{F} = \{(p, C) \in \mathcal{P} \times \mathfrak{C} \mid p \in C\}$, *see Lemma 2.1. Thus, we can use automorphisms of* $\Sigma(\mathcal{L}, \mathcal{R})$ *to map the subspace defined by the Jordan-system* $\mathcal{I}$ *elsewhere.*

Without loss of generality let $e_1^2 \neq 0$. We use the matrix $\begin{pmatrix} e_1 & 0 \\ 0 & 1 \end{pmatrix} = \gamma \in$ PGL($\mathcal{C}\ell_{(p,q,r)}, 2$) that corresponds to a right-multiplication with e_1 together with Lemma 2.1 and apply γ to $\mathcal{I}$. This results in a new subset that defines a chain geometry and is given by $\mathcal{J} = \mathcal{I}^\gamma =$span$(e_1, \ldots, e_n)$. Therefore, $\mathcal{I}$ is isomorphic to a Jordan-system $\mathcal{J}$ and the isomorphism is given by γ.

Remark 2.19. *For $\bigwedge^{n-1} V$ the statement also holds, since we can write a multiplication with the pseudoscalar as mapping $\gamma \in$ PGL($\mathcal{C}\ell_{(p,q,r)}, 2$). For subspaces $\bigwedge^k V$ with $1 < k < n-1$ it is not always possible to find a Jordan-system that can be mapped to these subspaces, but there are cases where it is possible. We proof that certain subspaces are isomorphic to Jordan-systems, when we need them.*

Remark 2.20. *A subspace $S \subseteq \bigwedge^1 V$ of a Clifford algebra over $\mathbb{R}$ is also isomorphic to a Jordan-closed strong Jordan-system.*

2.7.2 Pin and Spin Groups

Now we show that the Pin and the Spin group of a Clifford algebra correspond to a subspace.

Lemma 2.5. *The connected components of the Pin and the Spin group of a Clifford algebra satisfy the properties **J1-J3**, but they form no subspaces of the Clifford algebra $\mathcal{C}\ell_{(p,q,r)}$.*

Proof. In order to proof this theorem we have to check the condition **J1-J3** from the Def. 2.20 of a Jordan-system.

J1 Since the Pin and the Spin group are groups with respect to the Clifford product, the inverse elements are also contained in the group. Furthermore, every element possesses an inverse element and we have $\mathcal{J} = \mathcal{J}^\times = Pin^{\pm}_{(p,q,r)}$. We infer $\mathfrak{g} \in \mathcal{J} \Rightarrow \mathfrak{g}^{-1} \in \mathcal{J}$.

J2 The Pin respectively Spin group is closed under the Clifford product. Thus, for all $\mathfrak{a}, \mathfrak{b} \in Pin_{(p,q,r)} \Rightarrow \mathfrak{a}\mathfrak{b}\mathfrak{a} \in Pin_{(p,q,r)}$. Moreover, the norm is not changed, because the number of factors is odd.

J3 We show $|e(x)| > |\mathbb{R}\backslash e(x)|$. The set $\mathbb{R}\backslash e(x)$ is the set of all elements $r \in \mathbb{R}$ with $\mathfrak{a} - r \in \mathcal{N}, \mathfrak{a} \in Pin_{(p,q,r)}$. An element $\mathfrak{g}$ of a Clifford algebra is a zero divisor if $\mathfrak{g}\mathfrak{g}^* = 0$. Hence, we look at the equation

$$(\mathfrak{a} - r)(\mathfrak{a} - r)^* = 0.$$

This defines a quadratic equation in r with two real solutions at the most, and therefore, we have $|\mathbb{R}\backslash e(x)| = 2$. Thus, condition **J3** is satisfied.

We discussed the Pin group case, but for the Spin group the same arguments hold. $\square$

Lemma 2.5 encourages the assumption that subsets of the projective line $\mathbb{P}^1(\mathcal{Cl}_{(p,q,r)})$ defined by $\mathcal{Cl}^\times_{(p,q,r)}(\mathfrak{g},1)$, $\mathfrak{g} \in Pin^+_{(p,q,r)}$ and $\mathcal{Cl}^\times_{(p,q,r)}(\mathfrak{g},1)$, $\mathfrak{g} \in Pin^-_{(p,q,r)}$ are images of strong Jordan-closed Jordan-systems under a map $\gamma \in \mathrm{PGL}(\mathcal{Cl}_{(p,q,r)}, 2)$. To show that the connected components of the Pin group define chain geometries, we show the theorem:

Theorem 2.18. *A chain $c \in \mathfrak{C}$ defined by three points of the form $\mathcal{Cl}^\times_{(p,q,r)}(\mathfrak{g},1)$, $\mathfrak{g} \in Pin^\pm_{(p,q,r)}$ is completely contained in the embedding of $Pin^\pm_{(p,q,r)}$ in the projective line $\mathbb{P}^1(\mathcal{Cl}_{(p,q,r)})$.*

Proof. We define a mapping $\gamma \in \mathrm{PGL}(\mathcal{R}, 2)$ with $\mathcal{R}(1,0) \mapsto \mathcal{R}(\mathfrak{a},1)$, $\mathcal{R}(1,1) \mapsto \mathcal{R}(\mathfrak{b},1)$, and $\mathcal{R}(0,1) \mapsto \mathcal{R}(\mathfrak{c},1)$. Since the group $\mathrm{PGL}(\mathcal{R},2)$ acts flag-transitive on the set $\mathfrak{F}$, the standard chain $\mathcal{R}(s,r)$, $s,r \in \mathbb{R}$ is mapped to another chain. Thus, we have to show that every element of the chain $\mathbb{P}^1(\mathbb{R})^\gamma$ corresponds to an element $\mathcal{R}(\mathfrak{g},1)$ with $\mathfrak{g} \in Pin^\pm_{(p,q,r)}$. Therefore, we give the map γ in matrix form

$$M = \begin{pmatrix} v^{-1}\mathfrak{c} & v^{-1} \\ 1 - t\mathfrak{c} & -t \end{pmatrix} \text{ with } t = (\mathfrak{c} - \mathfrak{a})^{-1}, \ v = (t + (\mathfrak{b} - \mathfrak{c})^{-1})^{-1}.$$

Now, we apply this mapping to the standard chain and find

$$\mathcal{R}(s,r)M = \mathcal{R}(sv^{-1}\mathfrak{c} + r(1 - t\mathfrak{c}), sv^{-1} - tr) \sim \mathcal{R}(\mathfrak{c} + r(sv^{-1} - rt)^{-1}, 1).$$

Thus, we have to show that $\mathfrak{c} + r(sv^{-1} - rt)^{-1} \in Pin^\pm_{(p,q,r)}$ for all $\mathfrak{a}, \mathfrak{b}, \mathfrak{c} \in Pin^\pm_{(p,q,r)}$. We denote the group that fixes the set $\mathbb{P}^1(Pin^\pm_{(p,q,r)})$ by $\Delta(Pin^\pm_{(p,q,r)})$. The group $\Delta(Pin^\pm_{(p,q,r)})$ is generated by the matrices

$\begin{pmatrix} 0 & 1 \\ 1 & 0 \end{pmatrix}$, $\begin{pmatrix} \mathfrak{g} & 0 \\ 0 & 1 \end{pmatrix}$, and $\begin{pmatrix} 1 & 0 \\ 0 & \mathfrak{g} \end{pmatrix}$ with $\mathfrak{g} \in \mathrm{Pin}^{\pm}_{(p,q,r)}$. This group acts transitive on $\mathbb{P}^1(\mathrm{Pin}^{\pm}_{(p,q,r)})$, and therefore, we can assume that $\mathfrak{a} = 1$. To show that

$$(\mathfrak{c} + r(s v^{-1} - rt)^{-1})(\mathfrak{c} + r(s v^{-1} - rt)^{-1})^* = \pm 1,$$

we used the Clifford package from Maple, see [1, 2]. □

Remark 2.21. *The proof gives a new parametrisation of the chain*

$$c_{\mathfrak{abc}} = \mathfrak{c} + r(sv^{-1} - rt)^{-1}, \ r,s \in \mathbb{R}.$$

With Th. 2.18 each connected component of the Pin group defines a weak subspace and with Th. 2.13 a subspace of $\Sigma(\mathbb{R}, \mathcal{Cl}_{(p,q,r)})$.

Remark 2.22. *Th. 2.18 also holds for null vectors. Thus, the null vectors*

$$\mathcal{N}^1 := \left\{ \mathfrak{v} \in \bigwedge\nolimits^1 V \mid \mathfrak{v}\mathfrak{v} = 0 \right\}$$

are the image of a Jordan-system under a mapping $\gamma \in \mathrm{PGL}(\mathcal{Cl}_{(p,q,r)}, 2)$.

Remark 2.23. *It is obvious that also the restriction of subgroups of the Pin and the Spin group to one connected component are images of Jordan-systems, and therefore, chain geometries on their own. With [16] every Lie group can be represented as Spin group. Hence, we have a lot of examples for these chain spaces. Furthermore, Lie groups correspond to Lie algebras and subgroups of Lie groups to subalgebras of Lie algberas. The Lie algebra of a Spin group is given by its bivector algebra, see [17, p. 402].*

Now we give an example of a chain given by three elements of a Pin group. Therefore, we derive an explicit formula for the chain with Eq. (2.7). We can use this equation since all occurring algebra elements are proper elements. Let $\lambda \in \mathbb{R}$ be the cross ratio of four given elements. With Eq. (2.8) we get:

$$c_{\mathfrak{abc}}(\lambda) = \left[\mathfrak{a}(\mathfrak{b} - \mathfrak{a})^{-1}(\mathfrak{b} - \mathfrak{c}) - \lambda \mathfrak{c}\right] \left[(\mathfrak{b} - \mathfrak{a})^{-1}(\mathfrak{b} - \mathfrak{c}) - \lambda\right]^{-1}.$$

This gives a parametrisation of the chain $c_{\mathfrak{abc}}$ defined by the three points $\mathfrak{a}, \mathfrak{b}, \mathfrak{c} \in \mathrm{Pin}^{+}_{(p,q,r)}$ and the parameter equals the cross ratio.

2.7 Chain Geometries over Clifford Algebras

Example 2.2. *As example we look at the unit quaternions, that can be represented as Pin group elements, see Ex. 1.1. Let*

$$\mathfrak{a} = \frac{1}{\sqrt{2}}e_0 + \frac{1}{\sqrt{2}}e_{12}, \quad \mathfrak{b} = e_1, \quad \mathfrak{c} = e_2$$

be three algebra elements with $\mathfrak{aa}^* = \mathfrak{bb}^* = \mathfrak{cc}^* = 1$ of the Pin group of $\mathcal{Cl}_{(0,2,0)}$. The chain defined by these three elements parametrised with a real cross ratio is given by

$$c_{\mathfrak{abc}}(\lambda) = \frac{1}{2\lambda^2 - 2\lambda + 2}\left(\sqrt{2}(1-\lambda)e_0 + 2\lambda e_1 + 2\lambda(\lambda-1)e_2 + \sqrt{2}(1-\lambda)e_{12}\right).$$

We can check, that the chain is entirely contained in the Pin group

$$c_{\mathfrak{abc}}(\lambda)c_{\mathfrak{abc}}(\lambda)^* = 1, \quad \text{for all } \lambda \in \mathbb{R}.$$

Remark 2.24. *Since the group $\mathrm{PGL}(\mathcal{R}, 2)$ acts flag-transitive on $\mathfrak{F}$, we can use automorphisms of $\Sigma(\mathcal{L}, \mathcal{R})$ to map the projective line over the Pin or Spin group elsewhere.*

Example 2.3. *As an example we take the Clifford algebra $\mathcal{Cl}_{(0,2,0)} \cong \mathbb{H}$. The elements*

$$\mathfrak{a} = \sqrt{2}e_0 + \sqrt{2}e_{12}, \quad \mathfrak{b} = 2e_1, \quad \mathfrak{c} = 2e_2$$

are the elements of Ex. 2.2 multiplied by 2. Note, that the multiplication by 2 can also be expressed as mapping $\gamma \in \mathrm{PGL}(2, \mathcal{Cl}_{(0,2,0)})$. Now the elements $\mathfrak{a}, \mathfrak{b}, \mathfrak{c}$ satisfy $\mathfrak{aa}^* = \mathfrak{bb}^* = \mathfrak{cc}^* = 4$. Again we compute the chain with Eq. (2.8)

$$c_{\mathfrak{abc}}(\lambda) = \frac{1}{\lambda^2 - \lambda + 1}\left(\sqrt{2}(1-\lambda)e_0 + 2\lambda e_1 + 2\lambda(\lambda-1)e_2 + \sqrt{2}(1-\lambda)e_{12}\right).$$

This is exactly the chain of Ex. 2.2 multiplied by 2. Again all points of $c_{\mathfrak{abc}}(\lambda)$ satisfy

$$c_{\mathfrak{abc}}(\lambda)c_{\mathfrak{abc}}(\lambda)^* = 4, \quad \text{for all } \lambda \in \mathbb{R}.$$

Especially for the case of dual unit quaternions $\mathbb{U}_d \cong \mathrm{Spin}_{(3,0,1)}$ we are interested in the set of chains. Now we give an example for a strong Jordan-closed Jordan-system of the quaternions and construct a mapping that maps this Jordan-system to the Pin group.

Example 2.4. *A general quaternion or an element of $\mathcal{C}\ell_{(0,2,0)}$ has the form*

$$\mathfrak{q} = q_0 e_0 + q_1 e_1 + q_2 e_2 + q_{12} e_{12}.$$

*The set $\mathcal{J} = \{q_0 e_0 + q_1 e_1 + q_2 e_2 \mid q_0, q_1, q_2 \in \mathbb{R}\}$ forms a strong and Jordan-closed Jordan-system. We check condition **J1-J3**.*

J1 *Let $\mathfrak{q} = q_0 e_0 + q_1 e_1 + q_2 e_2 \in \mathcal{J}$, then its inverse is computed as*

$$\mathfrak{q}^{-1} = \frac{1}{q_0^2 + q_1^2 + q_2^2}(q_0 e_0 - q_1 e_1 - q_2 e_2) \in \mathcal{J}.$$

J2 *Let $\mathfrak{p} = p_0 e_0 + p_1 e_1 + p_2 e_2$ and $\mathfrak{q} = q_0 e_0 + q_1 e_1 + q_2 e_2$ be two elements of $\mathcal{J}$, then*

$$\mathfrak{qpq} = (q_0^2 p_0 - 2q_0 q_1 p_1 - 2q_0 q_2 p_2 - q_1^2 p_0 - q_2^2 p_0) e_0$$
$$+ (q_0^2 p_1 - 2q_1 q_2 p_2 - q_1^2 p_1 + q_2^2 p_1 + 2q_0 q_1 p_0) e_1$$
$$+ (2q_0 q_2 p_0 + q_0^2 p_2 + q_1^2 p_2 - q_2^2 p_2 - 2q_1 q_2 p_1) e_2 \in \mathcal{J}.$$

J3 *Quaternions, and therefore, the Clifford algebra contains no zero divisors. Thus, we infer*

$$|e(\mathfrak{q})| > |e(\mathbb{R}\backslash\mathfrak{q})|, \ \forall \mathfrak{q} \in \mathcal{J}.$$

Let $\mathfrak{a} = e_0$, $\mathfrak{b} = e_1$ and $\mathfrak{c} = \frac{1}{\sqrt{2}} e_2 + \frac{1}{\sqrt{2}} e_{12}$. We construct the mapping $\gamma \in \mathrm{PGL}(\mathcal{C}\ell_{(0,2,0)}, 2)$ that maps the points $\mathcal{R}(1,0) \mapsto \mathcal{R}(\mathfrak{a}, 1)$, $\mathcal{R}(1,1) \mapsto \mathcal{R}(\mathfrak{b}, 1)$ and $\mathcal{R}(0,1) \mapsto \mathcal{R}(\mathfrak{c}, 1)$ and apply it to $\mathcal{J}$. The matrix form of γ can be derived as

$$\mathrm{M} = \begin{pmatrix} v^{-1}\mathfrak{c} & v^{-1} \\ 1 - t\mathfrak{c} & -t \end{pmatrix}, \text{ with } t = (\mathfrak{c} - \mathfrak{a})^{-1}, \ v = (t + (\mathfrak{b} - \mathfrak{c})^{-1})^{-1},$$

cf. proof of Th. 2.18. We apply this mapping to a general element $\mathfrak{q} = q_0 e_1 + q_1 e_1 + q_2 e_2 \in \mathcal{J}$:

$$\mathcal{R}(\mathfrak{q}, 1)\mathrm{M} = \mathcal{R}(\mathfrak{q}v^{-1}\mathfrak{c} + (1 - t\mathfrak{c}), \mathfrak{q}v^{-1} - t) \sim \mathcal{R}(\mathfrak{c} + (\mathfrak{q}v^{-1} - t)^{-1}, 1).$$

The image of $\mathcal{J}$ can be described by

2.8 Quadric Chain Geometry

$$\mathfrak{c}+(\mathfrak{q}v^{-1}-t)^{-1} = \frac{1}{1+q_1-q_0+q_2^2+q_1^2+q_0^2}\Big((1-q_0+q_1)e_0 \qquad (2.11)$$
$$+ (q_0+q_1)e_1 + \frac{\sqrt{2}}{2}(q_1-q_0+q_2^2+q_1^2+q_0^2+\sqrt{2}q_2)e_2$$
$$+ \frac{\sqrt{2}}{2}(q_1-q_0+q_2^2+q_1^2+q_0^2-\sqrt{2}q_2)e_{12}\Big).$$

Every element of $\mathcal{J}^\gamma$ has norm equal to one and corresponds to an unit quaternion. This can be checked easily with Eq. 2.11:

$$\left(\mathfrak{c}+(\mathfrak{q}v^{-1}-t)^{-1}\right)\left(\mathfrak{c}+(\mathfrak{q}v^{-1}-t)^{-1}\right)^* = 1, \ \forall \mathfrak{q} \in \mathcal{J}.$$

2.8 Quadric Chain Geometry

A. BLUNCK showed in [10] that chain spaces on quadrics can be represented with the help of Jordan-systems $\mathcal{J}$ contained in a Clifford algebra. With the prerequisites from section 2.7.1 we are able to construct chain geometries corresponding to quadrics in a surprisingly simple way. For this purpose, we consider a quadric $\mathfrak{Q} \subseteq \mathbb{P}^n(\mathbb{R})$ given by the symmetric matrix $Q \in \mathbb{R}^{(n+1)\times(n+1)}$ corresponding to the quadratic form. To describe the quadric chain geometry algebraically we construct the Clifford algebra $\mathcal{C}\ell_{(p,q,r)}$ corresponding to $\mathfrak{Q}$. The points of the quadric can be represented as null vectors $\mathfrak{v} \in \bigwedge^1 V$ with $\mathfrak{v}\mathfrak{v} = 0$. So far we have collected all the results and settings that we need in order to construct the quadric chain geometry. In the following we give some examples to show the elegance of this apparatus. In section 2.12 we use this method to construct and describe biarcs on arbitrary quadrics.

Remark 2.25. *Since quadrics can be constructed over any commutative ring, we can generalize this method easily to quadric chain geometries over commutative rings, see [9].*

2.8.1 Klein's Quadric

As a first example of the algebraic description of a quadric chain geometry we consider Klein's quadric. We compute a regular conic defined by three points with methods of projective geometry. After that we compute the chain defined by these three points described as null vectors in the related Clifford algebra. With corresponding parametrisations we see that both methods result in exactly the same conic.

Example 2.5. *Let $A = a\mathbb{R} = (1,0,0,0,1,1)^T\mathbb{R}$, $B = b\mathbb{R} = (0,1,0,0,0,1)^T\mathbb{R}$, and $C = c\mathbb{R} = (0,0,1,1,0,0)^T\mathbb{R}$ be Plücker coordinates of three given lines, see Def. 1.4. Plücker coordinates satisfy the quadratic equation, cf. Eq. 1.12, and describe the points contained in a quadric M_2^4. The corresponding lines are visualized in Fig. 2.1. The quadratic form Q corresponding to M_2^4 is given by*

$$Q = \begin{pmatrix} O & I \\ I & O \end{pmatrix},$$

where I is the 3×3-identity matrix and O the 3×3 zero matrix. We denote the corresponding bilinear form by Ω. A general parametrisation of the conic section of M_2^4 with the two-space defined by A, B, C is given by

$$c(t) = t\ \Omega(A,B)C + (1-t)\ \Omega(A,C)B + t(t-1)\ \Omega(B,C)A. \quad (2.12)$$

Note, that the points A, B, and C correspond to the parameter values $t = \infty, 0, 1$. For the given points A, B, C the parametrisation (2.12) becomes

$$c(t) = (t(t-1), 2(1-t), t, t, t(t-1), t^2 - 3t + 2)^T. \quad (2.13)$$

Now we consider the Clifford algebra $\mathcal{Cl}_{(3,3,0)}$ defined by $\mathbb{R}^6$ as vector space model for $\mathbb{P}^5(\mathbb{R})$ and the quadratic form Q, see also section 1.6 or [40]. The null vectors corresponding to the points A, B, and C are given by

$$\mathfrak{a} = e_1 + e_5 + e_6, \quad \mathfrak{b} = e_2 + e_6, \quad \mathfrak{c} = e_3 + e_5.$$

If we use Eq. (2.8) the corresponding chain parametrised with a real cross ratio $\lambda \in \mathbb{R}$ is given by

$$c_{\mathfrak{abc}}(\lambda) = \frac{1}{\lambda^2 - 2\lambda + 2}(2(1-\lambda)e_1 + \lambda e_2 + \lambda(\lambda-1)e_3 \\ + (\lambda^2 - 3\lambda + 2)e_5 + (2-\lambda)e_6).$$

2.8 Quadric Chain Geometry

Since we are computing with homogeneous coordinates we can normalize to

$$c_{\mathfrak{abc}}(\lambda) = 2(1-\lambda)e_1 + \lambda e_2 + \lambda(\lambda-1)e_3 + (\lambda^2 - 3\lambda + 2)e_5 + (2-\lambda)e_6. \quad (2.14)$$

Equating the coefficients of Eq. (2.13) and Eq. (2.14) shows that both representations describe the same set of points if we identify $\bigwedge^1 V$ with $\mathbb{P}^5(\mathbb{R})$. Furthermore, the chain interpolates the null vectors $\mathfrak{a}, \mathfrak{b}, \mathfrak{c}$ for the cross ratio values $\lambda = \infty, 0, 1$. A visualization of the resulting regulus together with its striction curve (blue) is shown in Fig. 2.1.

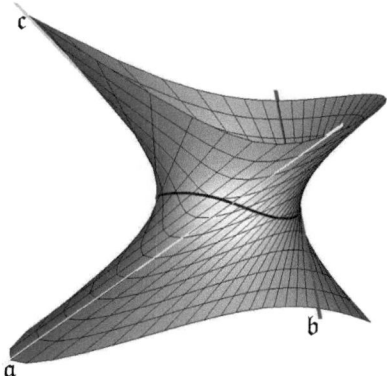

Figure 2.1: Regulus corresponding to Eq. (2.13) and Eq. (2.14)

2.8.2 Lie's Quadric

Lie's quadric was introduced in section 1.7. Now we use the homogeneous Clifford algebra model corresponding to Lie's quadric to describe its quadric chain geometry algebraically. The Clifford algebra model is given by the vector space $V = \mathbb{R}^6$ as an algebraic model of $\mathbb{P}^5(\mathbb{R})$ and the symmetric matrix of the quadratic form

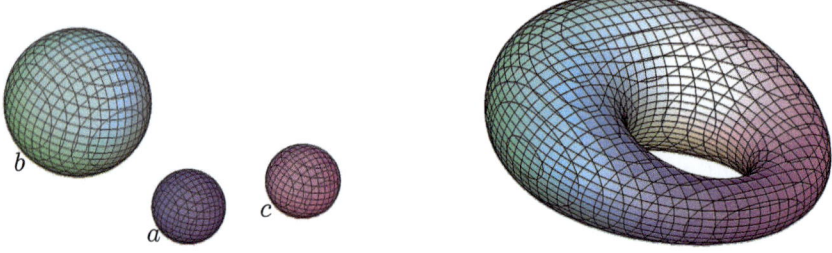

Figure 2.2: Dupin cyclide corresponding to the envelope of a conic on Lie's quadric

$$Q = \begin{pmatrix} -1 & 0 & 0 & 0 & 0 & 0 \\ 0 & 1 & 0 & 0 & 0 & 0 \\ 0 & 0 & 1 & 0 & 0 & 0 \\ 0 & 0 & 0 & 1 & 0 & 0 \\ 0 & 0 & 0 & 0 & 1 & 0 \\ 0 & 0 & 0 & 0 & 0 & -1 \end{pmatrix},$$

compare to section 1.7.2. The subspace $\bigwedge^1 V$ corresponds to the projective space $\mathbb{P}^5(\mathbb{R})$. Furthermore, the points contained on Lie's quadric are null vectors in the homogeneous Clifford algebra model.

Example 2.6. Let $A = a\mathbb{R} = (0, 1, 0, 0, 0, 1)^T \mathbb{R}$, $C = c\mathbb{R} = (8, -7, 4, 0, 0, 1)^T \mathbb{R}$, and $B = b\mathbb{R} = (11, -10, 0, 5, 0, 2)^T \mathbb{R}$ be three points on L_1^4. According to section 1.7.1 we interpret these points as Euclidean entities. The point A corresponds to a sphere with center $c = (0, 0, 0)^T$ and signed radius $r = 1$. Point B represents a sphere with center $c = (0, 5, 0)^T$ and signed radius $r = 2$. Further, C represents a sphere with center $c = (4, 0, 0)^T$ and $r = 1$. A visualization of the three spheres corresponding to A, B and C is given in Fig. 2.2 (left). We rewrite the points as null vectors

$$\mathfrak{a} = e_2 + e_6, \quad \mathfrak{b} = 11e_1 - 10e_2 + 5e_4 + 2e_6, \quad \mathfrak{c} = 8e_1 - 7e_2 + 4e_3 + e_6.$$

The chain defined by $\mathfrak{a}, \mathfrak{b}$, and $\mathfrak{c}$ is computed with Eq. (2.8):

$$c_{\mathfrak{abc}}(\lambda) = \frac{1}{5\lambda^2 - 4\lambda + 2}\left(2(\lambda + 11)e_1 + (5\lambda^2 - 6\lambda - 20)e_2 + 12\lambda e_3\right.$$

2.8 Quadric Chain Geometry

$$+ 10(1-\lambda)e_4 + (5\lambda^2 - 6\lambda + 4)e_6).$$

Note, that this parametrisation satisfies $x_1 + x_2 = 1$, where x_1 and x_2 are the coefficients of e_1 respectively e_2. Thus, we can read off the curve of centers of the oriented spheres

$$c(\lambda) = \frac{1}{5\lambda^2 - 4\lambda + 2} (12\lambda, 10(1-\lambda), 0)^\mathrm{T}$$

and the radius function as

$$r(\lambda) = \frac{5\lambda^2 - 6\lambda + 4}{5\lambda^2 - 4\lambda + 2}.$$

It is well-known that conics on Lie's quadric correspond to a family of spheres whose envelope is a Dupin cyclide. The envelope of this one-parameter family of spheres is given by the implicit equation

$$\begin{aligned}0 =& 3(x^4 + y^4 + z^4) - 30y^3 - 24x^3 + (36 + 6y^2 + 6z^2 - 30y)x^2 \\&+ (120y - 24y^2 - 24z^2 + 48)x + (6z^2 + 58)y^2 + (10 - 30z^2)y - 113 + 108z^2\end{aligned}$$

and visualized in Fig. 2.2 (right).

2.8.3 Study's Quadric

The homogeneous model corresponding to Study's quadric was introduced in section 1.8. It is constructed with the eight-dimensional vector space $\mathbb{R}^8$ as model for $\mathbb{P}^7(\mathbb{R})$ and the quadratic form

$$Q = \begin{pmatrix} O & I \\ I & O \end{pmatrix},$$

where O is the 4×4 zero matrix and I is the 4×4 identity matrix. Similar to the examples for Klein's and Lie's quadric we give an example of a conic on Study's quadric as chain in the homogeneous Clifford algebra model.

Example 2.7. Let $A = (1, 0, 0, 0, 0, 1, 1, 1)^\mathrm{T}\mathbb{R}$, $B = (0, 1, 0, 0, 0, 0, 1, 1)^\mathrm{T}\mathbb{R}$ and $C = (0, 0, 1, 0, 1, 0, 0, 0)^\mathrm{T}\mathbb{R}$ be three points on $S_2^6 \subseteq \mathbb{P}^7(\mathbb{R})$. In analogy to Eq. (2.12) the conic that interpolates A, B, and C for the parameter values $t = \infty, 0$, and 1 can be parametrised by

$$c(t) = t\Omega(A,B)C + (1-t)\Omega(A,C)B + t(t-1)\Omega(B,C)A,$$

where Ω denotes the bilinear form belonging to Study's quadric. Especially for the chosen points we derive:

$$c(t)=t\begin{pmatrix}0\\0\\1\\0\\1\\0\\0\\0\end{pmatrix}+2(1-t)\begin{pmatrix}0\\1\\0\\0\\0\\0\\0\\1\end{pmatrix}+t(t-1)\begin{pmatrix}1\\0\\0\\0\\0\\1\\1\\1\end{pmatrix}=\begin{pmatrix}t(t-1)\\2(1-t)\\t\\0\\t\\t(t-1)\\t^2-3t+2\\t^2-3t+2\end{pmatrix}.$$

Now we calculate a parametrisation of the chain that is defined by A, B, and C if we identify $\mathbb{P}^7(\mathbb{R})$ with $\bigwedge^1 V$. The corresponding null vectors have the form

$$\mathfrak{a} = e_1+e_6+e_7+e_8, \quad \mathfrak{b} = e_2+e_7+e_8, \quad \mathfrak{c} = e_3+e_5.$$

The chain defined by these three null vectors is computed with Eq. (2.8)

$$c_{\mathfrak{abc}}(\lambda) = \frac{1}{\lambda^2-2\lambda+2}(\lambda(\lambda-1)e_1-2(\lambda-1)e_2+\lambda e_3$$
$$+\lambda e_5+\lambda(\lambda-1)e_6+(\lambda^2-3\lambda+2)e_7+(\lambda^2-3\lambda+2)e_8).$$

Normalizing this equation results in

$$c_{\mathfrak{abc}}(\lambda) = \lambda(\lambda-1)e_1-2(\lambda-1)e_2+\lambda e_3+\lambda e_5$$
$$+\lambda(\lambda-1)e_6+(\lambda^2-3\lambda+2)e_7+(\lambda^2-3\lambda+2)e_8.$$

Equating the coefficients shows that both representations are exactly the same. The chain $c_{\mathfrak{abc}}$ is contained in the two-space spanned by the points A, B, C that correspond to the null vectors $\mathfrak{a}, \mathfrak{b}, \mathfrak{c}$.

2.8.4 Study's Sphere

The dual sphere as point model for oriented lines was introduced in section 1.2.3. Now we aim at a Clifford algebra that embeds Study's sphere in the grade-1 subspace $\bigwedge^1 V$ in a natural way. Therefore,

2.8 Quadric Chain Geometry

we construct the Clifford algebra model over the module $\mathcal{M} = \mathbb{D}^3$ together with the quadratic form

$$Q = \begin{pmatrix} 1 & 0 & 0 \\ 0 & 1 & 0 \\ 0 & 0 & 1 \end{pmatrix}.$$

For a general grade-1 element $\mathfrak{v} = x_1 e_1 + x_2 e_2 + x_3 e_3$ with $x_1 = a_1 + \epsilon b_1$, $x_2 = a_2 + \epsilon b_2$ and $x_3 = a_3 + \epsilon b_3$ we get

$$\mathfrak{v}\mathfrak{v} = x_1^2 + x_2^2 + x_3^2 = a_1^2 + a_2^2 + a_3^2 + 2\epsilon(a_1 b_1 + a_2 b_2 + a_3 b_3).$$

Since we describe the dual unit sphere, the real part has to be equal to 1 and dual part has to vanish. We already showed that the connected components of the Pin group define subspaces of $\Sigma(\mathcal{C}(\mathcal{Cl}_{(p,q,r)}), \mathcal{Cl}_{(p,q,r)})$, see section 2.7.2. Note that we use an affine Clifford algebra model here. It is also possible to construct a homogeneous. We perform this construction for the case of dual unit quaternions.

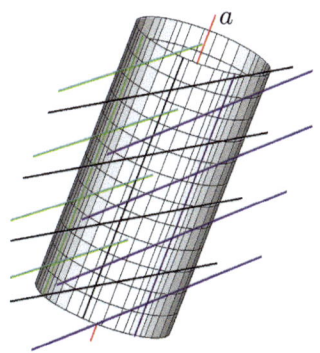

Figure 2.3: Line congruence with constant dual angle to an axis a (red)

Example 2.8. *Now we give an example for a chain that corresponds to a conic section on $S_{\mathbb{D}}^2$. Let $A = (1,0,0,0,1,1)^T \mathbb{R}$, $B = (0,1,0,0,0,1)^T \mathbb{R}$,* and $C = (0, 0, 1, 1, 0, 0)^T \mathbb{R}$ be three lines given in Plücker coordinates. These Plücker coordinates are the same that were used in Ex. 2.5. The corresponding dual vectors

$$\mathbf{a} = \begin{pmatrix} 1 \\ \epsilon \\ \epsilon \end{pmatrix}, \quad \mathbf{b} = \begin{pmatrix} 0 \\ 1 \\ \epsilon \end{pmatrix}, \quad \mathbf{c} = \begin{pmatrix} \epsilon \\ 0 \\ 1 \end{pmatrix}$$

are already normed. Written as algebra elements the three oriented lines get the form:

$$\mathfrak{a} = e_1 + \epsilon e_2 + \epsilon e_3, \quad \mathfrak{b} = e_2 + \epsilon e_3, \quad \mathfrak{c} = \epsilon e_1 + e_3.$$

Remember that the quadric is constructed over the ring of dual numbers $\mathbb{D}$. Therefore, the cross ratio may be a dual number $\Lambda = \lambda + \epsilon \mu$. The chain, computed with Eq. (2.8), has the form

$$c_{\mathfrak{abc}}(\Lambda) = c_{\mathfrak{abc}}(\lambda, \mu) = -\frac{1 - \lambda + \epsilon(\lambda^2 - \mu - 1)e_1}{(\lambda - \lambda^2 - 1 + \lambda^2 \epsilon - 2\mu\lambda\epsilon + \epsilon + \mu\epsilon)}$$
$$+ \frac{-\lambda + \epsilon(3\lambda - \mu - 1)e_2}{(\lambda - \lambda^2 - 1 + \lambda^2 \epsilon - 2\mu\lambda\epsilon + \epsilon + \mu\epsilon)}$$
$$+ \frac{-\lambda^2 + \lambda + \epsilon(\lambda^2 - 2\mu\lambda - \lambda + \mu - 1)e_3}{\lambda - \lambda^2 - 1 + \lambda^2 \epsilon - 2\mu\lambda\epsilon + \epsilon + \mu\epsilon)}.$$

We split the dual numbers into real and dual parts and get

$$c_{\mathfrak{abc}}(\lambda, \mu) = \left(\frac{1 - \lambda}{\lambda^2 - \lambda + 1} + \epsilon \frac{\lambda(\lambda^3 - 2\lambda^2 + \mu\lambda + \lambda - 2\mu)}{(\lambda^2 - \lambda + 1)^2} \right) e_1 \quad (2.15)$$
$$+ \left(\frac{\lambda}{\lambda^2 - \lambda + 1} + \epsilon \frac{(1 - \lambda)(2\lambda^2 - 2\lambda + \mu\lambda + 1 + \mu)}{(\lambda^2 - \lambda + 1)^2} \right) e_2$$
$$+ \left(\frac{(\lambda - 1)\lambda}{\lambda^2 - \lambda + 1} + \epsilon \frac{\lambda^3 - \lambda + 1 - \mu + 2\mu\lambda}{(\lambda^2 - \lambda + 1)^2} \right) e_3.$$

STUDY [61] and KLEIN [43] already knew that the intersection of the dual sphere $S_\mathbb{D}^2$ with two-dimensional affine submodules results in congruences of lines, i.e., two-parameter families of oriented lines that enclose a constant dual angle with an unique oriented line. Therefore, Eq. (2.15) is a parametrisation for this congruence of lines, see Fig. 2.3 for a schematic view. This congruence can be interpreted as the set of all tangents of a cylinder of revolution that enclose a constant angle with its axis.

2.9 Quadric Chain Spaces for certain Spin Groups

We use the work of BLUNCK [9] which shows that algebras of the form $\mathcal{L} + s\mathcal{L}$ with $s^2 \in \{1, -1, 0\}$ and $\mathcal{L}$ a local ring posses a quadric chain space over one of the rings $\mathbb{A}, \mathbb{C}$, or $\mathbb{D}$. Furthermore, we restrict ourselves to Spin group elements.

2.9.1 A Quadric Model for dual unit Quaternions

In this section we focus on the cross ratio of dual unit quaternions. We know that the dual unit quaternions are isomorphic to $\mathrm{Spin}_{(3,0,1)}$ and that this Spin group is contained in the even part of the algebra $\mathcal{Cl}^+_{(3,0,1)} \cong \mathcal{Cl}_{(2,0,1)}$. Furthermore, the algebra is of the form $\mathcal{L} + s\mathcal{L}$, with $s^2 = 0$, since $\mathcal{Cl}_{(2,0,1)} \cong \mathcal{Cl}_{(2,0,0)} \otimes \mathbb{D}$. For our investigations it is important that the norm of a general element $\mathfrak{A} \in \mathcal{Cl}^+_{(3,0,1)}$ is contained in the center $\mathcal{C}(\mathcal{Cl}^+_{(3,0,1)}) \cong \mathbb{D}$. Let

$$\mathfrak{g} = a_0 + a_3 e_{12} + a_2 e_{13} + c_1 e_{14} + a_1 e_{23} + c_2 e_{24} + c_3 e_{34} + c_0 e_{1234}$$

be a general element from $\mathcal{Cl}^+_{(3,0,1)}$. We calculate its norm by

$$\mathfrak{g}\mathfrak{g}^* = (a_0^2 + a_1^2 + a_2^2 + a_3^2) + 2(c_0 a_0 - a_1 c_1 + c_2 a_2 - c_3 a_3)e_{1234}. \quad (2.16)$$

Setting the pseudoscalar part of Eq. (2.16) to zero yields $c_0 a_0 - a_1 c_1 + c_2 a_2 - c_3 a_3 = 0$. This describes a quadric that is the image of Study's quadric under a collineation. This collineation defines a bijection between $\mathrm{Spin}_{(3,0,1)}$ and $\mathbb{U}_d$, see chapter 3. A spin group element $\mathfrak{g} \in \mathcal{Cl}^+_{(3,0,1)}$ satisfies

$$\mathfrak{g}\mathfrak{g}^* = (a_0^2 + a_1^2 + a_2^2 + a_3^2) + 2(c_0 a_0 - a_1 c_1 + c_2 a_2 - c_3 a_3)e_{1234} = 1.$$

Hence, we have the two conditions $a_0^2 + a_1^2 + a_2^2 + a_3^2 = 1$ and $c_0 a_0 - a_1 c_1 + c_2 a_2 - c_3 a_3 = 0$. In analogy to Ex. 2.8 we can construct an affine Clifford algebra model that describes the group $\mathrm{Spin}_{(3,0,1)}$ as the intersection of the grade-1 subspace with its Pin group.

The affine Clifford Algebra Model The corresponding affine model is constructed with the module $\mathcal{M} = \mathbb{D}^4$ and
$$Q = \begin{pmatrix} 1 & 0 & 0 & 0 \\ 0 & 1 & 0 & 0 \\ 0 & 0 & 1 & 0 \\ 0 & 0 & 0 & 1 \end{pmatrix}.$$

Hence, the square of grade-1 elements is the canonical scalar product of dual vectors. Let $\mathfrak{v} = (a_0+\epsilon c_0)e_1+(a_1+\epsilon c_1)e_2+(a_2+\epsilon c_2)e_3+(a_3+\epsilon c_3)e_4$ be a general element. we compute
$$\mathfrak{v}\mathfrak{v} = a_0^2 + a_1^2 + a_2^2 + a_3^2 + 2\epsilon(a_0c_0 + a_1c_1 + a_2c_2 + a_3c_3).$$

This is exactly the norm of a dual quaternion $q = a_0 + a_1\mathbf{i} + a_2\mathbf{j} + a_3\mathbf{k} + \epsilon(c_0 + c_1\mathbf{i} + c_2\mathbf{j} + c_3\mathbf{k})$. The grade-1 subspace is isomorphic to a Jordan-system, the Pin group is a subspace, and therefore, the restriction of the Pin-group to the grade-1 subspace is also a subspace. Thus, chains of dual quaternions with norm equal to 1 can be represented in $\mathrm{Pin}(\mathcal{Cl}_{(0,4,0)}(\mathbb{D})) \cap \bigwedge^1(\mathcal{Cl}_{(0,4,0)}(\mathbb{D}))$. The chains are conic sections of the dual sphere $S_\mathbb{D}^3 \subseteq \mathbb{D}^4$ with dual planes. A geometric interpretation of the corresponding displacements is given in section 2.10.

Remark 2.26. *The Study sphere $S_\mathbb{D}^2$ is contained in this model. If we set $a_0 = c_0 = 0$ the corresponding displacements are line-symmetric. They still satisfy the quadric equation and can be interpreted as oriented lines. Hence, the point model for oriented lines $S_\mathbb{D}^2$ is naturally embedded in $S_\mathbb{D}^3$.*

The homogeneous Clifford Algebra Model It is also possible to use a projective quadric and the corresponding quadratic form to define a Clifford algebra model. In this setting the interesting elements are embedded as null vectors. To construct the corresponding Clifford algebra model we use $\mathcal{M} = \mathbb{D}^5$ as a model for the projective space $\mathbb{P}^4(\mathbb{D})$ together with the quadratic form corresponding to a sphere
$$Q = \begin{pmatrix} -1 & 0 & 0 & 0 & 0 \\ 0 & 1 & 0 & 0 & 0 \\ 0 & 0 & 1 & 0 & 0 \\ 0 & 0 & 0 & 1 & 0 \\ 0 & 0 & 0 & 0 & 1 \end{pmatrix}.$$

The corresponding homogeneous Clifford algebra model is given by $\mathcal{C}\ell_{(1,4,0)}(\mathbb{D})$.

Remark 2.27. *For $n = p+q+r = 4$ we have some more cases that allow a description as quadric model over a commutative ring.*

2.9.2 Other possible Quadric Models

$\mathcal{C}\ell_{(4,0,0)}$: We start with the Clifford algebra $\mathcal{C}\ell_{(4,0,0)}$. Its Spin group is a double cover of the group SO(4). A general element of the even part has the form

$$\mathfrak{g} = a_0 + a_3 e_{12} + a_2 e_{13} + c_1 e_{14} + a_1 e_{23} + c_2 e_{24} + c_3 e_{34} + c_0 e_{1234}.$$

The product with its conjugate element is calculated as

$$\mathfrak{g}\mathfrak{g}^* = (a_0^2 + a_1^2 + a_2^2 + a_3^2 + c_0^2 + c_1^2 + c_2^2 + c_3^2) e_0 + 2(c_0 a_0 - a_1 c_1 + c_2 a_2 - c_3 a_3) e_{1234}.$$

If this is an element of the Spin group, the real part has to equal 1 while the pseudoscalar part has to vanish. The equation in the pseudoscalar part can, again, be understood as the equation of Study's quadric. Therefore, the same set of points in projective space can serve as a point model for different geometric entities, see chapter 3.

Affine Clifford Algebra Model Since the center of $\mathcal{C}\ell_{(4,0,0)}$ is isomorphic to $\mathbb{A}$ we construct the affine model by the use of $\mathcal{M} = \mathbb{A}^4$ and the quadratic form

$$Q = \begin{pmatrix} 1 & 0 & 0 & 0 \\ 0 & 1 & 0 & 0 \\ 0 & 0 & 1 & 0 \\ 0 & 0 & 0 & 1 \end{pmatrix}.$$

The chain geometry of $\mathrm{Spin}_{(4,0,0)}$ can then be found in the grade-1 space. A general vector $\mathfrak{v} \in \bigwedge^1 V$ has the form $\mathfrak{v} = (a_0 + jc_0) e_1 + (a_1 - jc_1) e_2 + (a_2 + jc_2) e_3 + (a_3 - jc_3) e_4$ with $j^2 = 1$. Its square is

$$\mathfrak{v}\mathfrak{v} = a_0^2 + a_1^2 + a_2^2 + a_3^2 + c_0^2 + c_1^2 + c_2^2 + c_3^2 + 2j(c_0 a_0 - a_1 c_1 + c_2 a_2 - c_3 a_3).$$

This is exactly the norm of a general element from $\mathcal{C}\ell^+_{(4,0,0)}$. The chains are conic sections with split-complex planes, *i.e.*, two-dimensional affine submodules in the module $\mathbb{A}^4$.

Homogeneous Clifford Algebra Model To construct the homogeneous Clifford algebra model we use $\mathcal{M} = \mathbb{A}^5$ as model for the projective space $\mathbb{P}^4(\mathbb{A})$ and the quadratic form

$$Q = \begin{pmatrix} -1 & 0 & 0 & 0 & 0 \\ 0 & 1 & 0 & 0 & 0 \\ 0 & 0 & 1 & 0 & 0 \\ 0 & 0 & 0 & 1 & 0 \\ 0 & 0 & 0 & 0 & 1 \end{pmatrix}.$$

Like in the case of dual quaternions the unit double quaternions are in this model represented as null vectors.

$\mathcal{C}\ell_{(3,1,0)}$: For a geometric interpretation of this algebra and its Spin group, see chapter 3. We construct the quadric model for the Spin group in the same way. Thus, let

$$\mathfrak{g} = a_0 + a_3 e_{12} + a_2 e_{13} + c_1 e_{14} + a_1 e_{23} + c_2 e_{24} + c_3 e_{34} + c_0 e_{1234}$$

be an arbitrary element in $\mathcal{C}\ell_{(3,1,0)}$. The product with its conjugate element results in

$$\mathfrak{g}\mathfrak{g}^* = (a_0^2 + a_1^2 + a_2^2 + a_3^2 - c_0^2 - c_1^2 - c_2^2 - c_3^2)e_0 + 2(a_0 c_0 - a_1 c_1 + a_2 c_2 - a_3 c_3)e_{1234}.$$

Affine Clifford Algebra Model In order to construct the affine model we need $V = \mathbb{C}^4$ as vector space and

$$Q = \begin{pmatrix} 1 & 0 & 0 & 0 \\ 0 & 1 & 0 & 0 \\ 0 & 0 & 1 & 0 \\ 0 & 0 & 0 & 1 \end{pmatrix}$$

as quadratic form. A general vector $\mathfrak{v} \in \bigwedge^1 V$ has the form $\mathfrak{v} = (a_0 + ic_0)e_1 + (a_1 - ic_1)e_2 + (a_2 + ic_2)e_3 + (a_3 - ic_3)e_4$. Its square reads

$$\mathfrak{v}\mathfrak{v} = a_0^2 + a_1^2 + a_2^2 + a_3^2 - c_0^2 - c_1^2 - c_2^2 - c_3^2 + 2(c_0 a_0 - a_1 c_1 + c_2 a_2 - c_3 a_3)i.$$

This is exactly the norm of a general element of $\mathcal{C}\ell^+_{(3,1,0)}$.

2.9 Quadric Chain Spaces for certain Spin Groups

Homogeneous Clifford Algebra Model For the homogeneous quadric model of $\mathrm{Spin}_{(3,1,0)}$ we use $V = \mathbb{C}^5$ as vector space model for $\mathbb{P}^4(\mathbb{C})$. The corresponding quadratic form is given by

$$Q = \begin{pmatrix} -1 & 0 & 0 & 0 & 0 \\ 0 & 1 & 0 & 0 & 0 \\ 0 & 0 & 1 & 0 & 0 \\ 0 & 0 & 0 & 1 & 0 \\ 0 & 0 & 0 & 0 & 1 \end{pmatrix}.$$

Remark 2.28. *The Clifford algebra $C\ell_{(3,1,0)}$ results in the same model. We have to use the quadratic form $Q = \mathrm{diag}(1, 1, 1, -1)$ for $C\ell_{(3,1,0)}$. If we use the standard form $Q = \mathrm{diag}(-1, 1, 1, 1)$, the equality is not obvious.*

$C\ell_{(2,2,0)}$: A general element of the even part of $C\ell_{(2,2,0)}$ is given by

$$\mathfrak{g} = a_0 + a_3 e_{12} + a_2 e_{13} + c_1 e_{14} + a_1 e_{23} + c_2 e_{24} + c_3 e_{34} + c_0 e_{1234}.$$

Its norm is determined by

$$\mathfrak{g}\mathfrak{g}^* = (a_0^2 - a_1^2 - a_2^2 + a_3^2 + c_0^2 - c_1^2 - c_2^2 + c_3^2)e_0 + 2(a_0 c_0 - a_1 c_1 + a_2 c_2 - a_3 c_3)e_{1234}.$$

We proceed analogue to the cases above and determine the affine and the homogeneous Clifford algebra models for this Spin group.

Affine Clifford Algebra Model As module we use $\mathcal{M} = \mathbb{A}^4$. The corresponding quadratic form is computed as

$$Q = \begin{pmatrix} 1 & 0 & 0 & 0 \\ 0 & -1 & 0 & 0 \\ 0 & 0 & 1 & 0 \\ 0 & 0 & 0 & -1 \end{pmatrix}.$$

A general vector $\mathfrak{v} \in \bigwedge^1 V$ is given by $\mathfrak{v} = (a_0 + \sigma c_0)e_1 + (a_1 + \sigma c_1)e_2 + (a_3 + \sigma c_3)e_3 + (a_2 + \sigma c_2)e_4$. Its square equals

$$\mathfrak{v}\mathfrak{v} = a_0^2 - a_1^2 - a_2^2 + a_3^2 + c_0^2 - c_1^2 - c_2^2 + c_3^2 + 2\sigma(+c_0 a_0 - a_1 c_1 - c_2 a_2 + c_3 a_3).$$

Again, we get exactly the norm of a general element from $C\ell^+_{(2,2,0)}$.

Homogeneous Clifford Algebra Model To represent the points of the group $\mathrm{Spin}_{(2,2,0)}$ as null vectors we construct a homogeneous Clifford algebra model with $\mathcal{M} = \mathbb{A}^5$ and

$$Q = \begin{pmatrix} -1 & 0 & 0 & 0 & 0 \\ 0 & 1 & 0 & 0 & 0 \\ 0 & 0 & -1 & 0 & 0 \\ 0 & 0 & 0 & 1 & 0 \\ 0 & 0 & 0 & 0 & -1 \end{pmatrix}.$$

$\mathcal{Cl}_{(2,1,1)}$: A general element from the even part is again given by

$$\mathfrak{g} = a_0 + a_3 e_{12} + a_2 e_{13} + c_1 e_{14} + a_1 e_{23} + c_2 e_{24} + c_3 e_{34} + c_0 e_{1234}.$$

Its norm is calculated as

$$\mathfrak{g}\mathfrak{g}^* = (a_0^2 - a_1^2 - a_2^2 + a_3^2)e_0 + 2(c_0 a_0 - a_1 c_1 + c_2 a_2 - c_3 a_3)e_{1234}.$$

Affine Clifford Algebra Model The affine Clifford algebra is constructed with $\mathcal{M} = \mathbb{D}^4$ together with the quadratic form

$$Q = \begin{pmatrix} 1 & 0 & 0 & 0 \\ 0 & -1 & 0 & 0 \\ 0 & 0 & -1 & 0 \\ 0 & 0 & 0 & 1 \end{pmatrix}.$$

A vector has the form $\mathfrak{v} = (a_0 + \epsilon c_0)e_1 + (a_1 + \epsilon c_1)e_2 + (a_2 - \epsilon c_2)e_3 + (a_3 - \epsilon c_3)e_4$ and its square reads

$$\mathfrak{v}\mathfrak{v} = a_0^2 - a_1^2 - a_2^2 + a_3^2 + 2\epsilon(c_0 a_0 - a_1 c_1 + c_2 a_2 - c_3 a_3).$$

Again the chain geometry $\Sigma(\mathcal{C}(\mathcal{Cl}_{(2,1,0)}), \mathrm{Spin}_{(2,1,0)})$ is now represented as the set of conic sections respectively as chains in $\bigwedge^1 V \cap \mathrm{Pin}_{(2,1,0)}$.

Homogeneous Clifford Algebra Model To represent this quadric model in a homogeneous way, we embed this quadric in the projective space $\mathbb{P}^4(\mathbb{D})$. Thus, the homogeneous Clifford algebra model is constructed with $\mathcal{M} = \mathbb{D}^5$ as model for $\mathbb{P}^4(\mathbb{D})$ and the quadratic form

$$Q = \begin{pmatrix} -1 & 0 & 0 & 0 & 0 \\ 0 & 1 & 0 & 0 & 0 \\ 0 & 0 & -1 & 0 & 0 \\ 0 & 0 & 0 & -1 & 0 \\ 0 & 0 & 0 & 0 & 1 \end{pmatrix}.$$

Remark 2.29. *The Clifford algebra $C\ell_{(1,2,1)}$ results in a quadric with the same signature.*

Remark 2.30. *For Clifford algebras with more than one generator squaring to zero this construction does not work.*

2.10 Cross Ratio of dual unit Quaternions

In this section we focus on the question for the cross ratio of dual unit quaternions. To give an answer to this question we use a Lemma that is taken from [30].

Lemma 2.6. *Assume we are given three points $P_i \in S_2^6$, $i = 0, 1, 2$. Their Study coordinates are denoted by (a_i, c_i), where $a_i = (a_{i0}, a_{i1}, a_{i2}, a_{i3})^T$, $c_i = (c_{i0}, c_{i1}, c_{i2}, c_{i3})^T$, $i = 0, 1, 2$ are real vectors. Additionally, we assume*

$$rk \begin{pmatrix} a_0 & a_1 & a_2 \\ c_0 & c_1 & c_2 \end{pmatrix} = 3.$$

If $rk(a_0, a_1, a_2) = 3$, then an unique collineation κ, see Eq. (1.9), is determined such that any P_i^κ satisfies the equations $a_{i0} = c_{i0} = 0$.

A proof can be found in [30].

Remark 2.31. *With the help of Lemma 2.6 we can say: There exists exactly one coordinate transformation that maps three given displacements to line-symmetric displacements of the reference coordinate system.*

We do not need additional cases of Lemma 2.6, since a chain is defined by three mutually distant points, and therefore, $rk(a_0, a_1, a_2) = 3$ is always true for any three points of a chain. With the help of Lemma 2.6 we can restrict ourselves to line-symmetric motions. Now we show that the bivector part of $C\ell_{(3,0,1)}$ is isomorphic to a Jordan-closed strong Jordan-system.

Theorem 2.19. *The bivector subspace $\bigwedge^2 V$ of the Clifford algebra $C\ell_{(3,0,1)}$ is isomorphic to a Jordan-closed strong Jordan-system, i.e., $\Sigma(\mathcal{C}(C\ell_{(3,0,1)}), \bigwedge^2 V)$ is a subspace of $\Sigma(\mathcal{C}(C\ell_{(3,0,1)}), C\ell_{(3,0,1)})$.*

Proof. To proof this we show that the subspace $\mathcal{I}$ spanned by $\{e_0, e_{13}, e_{14}, e_{23}, e_{24}, e_{1234}\}$ is a Jordan-closed, strong Jordan-system. With the mapping $\begin{pmatrix} e_{12} & 0 \\ 0 & 1 \end{pmatrix} \in \mathrm{PGL}(\mathcal{C}\ell_{(3,0,1)}, 2)$ we can map the subspace to bivector subspace bijectively. Clearly $1 \in \mathcal{I}$. We show **J1-J3** from Def. 2.20:

J1 A general element from $\mathcal{I} \cap \mathcal{C}\ell_{(3,0,1)}^\times$ can be written as

$$\mathfrak{A} = x_1 e_0 - x_2 e_{23} + x_3 e_{13} - x_4 e_{24} + x_5 e_{14} - x_6 e_{1234}.$$

Its inverse element is computed as

$$\mathfrak{A}^{-1} = \frac{x_1}{x_1^2 + x_2^2 + x_3^2} e_0 + \frac{x_2}{x_1^2 + x_2^2 + x_3^2} e_{23} - \frac{x_3}{x_1^2 + x_2^2 + x_3^2} e_{13}$$
$$+ \frac{x_1^2 x_4 - 2x_3 x_1 x_6 + x_4 x_2^2 - x_3^2 x_4 + 2x_3 x_5 x_2}{(x_1^2 + x_2^2 + x_3^2)^2} e_{24}$$
$$- \frac{2x_2 x_6 x_1 + 2x_2 x_3 x_4 - x_2^2 x_5 + x_5 x_1^2 + x_3^2 x_5}{(x_1^2 + x_2^2 + x_3^2)^2} e_{14}$$
$$+ \frac{2x_3 x_1 x_4 + x_6 x_1^2 - x_3^2 x_6 - 2x_2 x_5 x_1 - x_2^2 x_6) e_{1234}}{(x_1^2 + x_2^2 + x_3^2)^2} \in \mathcal{I}.$$

It follows that $x_1^2 + x_2^2 + x_3^2 \neq 0$ for $\mathfrak{A} \in \mathfrak{A}^\times := \mathcal{I} \cap \mathcal{C}\ell_{(3,0,1)}^\times$.

J2 Let $\mathfrak{B} \in \mathcal{I}$ be another general element with

$$\mathfrak{B} = y_1 e_0 - y_2 e_{23} - y_3 e_{24} + y_4 e_{13} + y_5 e_{14} - y_6 e_{1234}.$$

We compute

$$\mathfrak{ABA} = (x_1^2 y_1 - 2x_3 x_1 y_4 - 2x_1 x_2 y_2 - x_2^2 y_1 - x_3^2 y_1) e_0$$
$$+ (x_2^2 y_4 + x_1^2 y_4 + 2x_3 x_1 y_1 - x_3^2 y_4 - 2x_2 x_3 y_2) e_{13}$$
$$- (-2x_2 x_3 y_4 + 2x_1 x_2 y_1 - x_2^2 y_2 + x_3^2 y_2 + x_1^2 y_2) e_{23}$$
$$+ (-2x_1 x_2 y_6 + x_3^2 y_5 + x_1^2 y_5 - 2x_5 x_2 y_2 - 2x_4 x_3 y_2 - 2x_5 x_3 y_4$$
$$- 2x_2 x_6 y_1 + 2x_1 x_5 y_1 - 2x_1 x_6 y_2 + 2x_2 x_3 y_3 + 2x_4 x_2 y_4 - x_2^2 y_5) e_{14}$$
$$- (2x_1 x_4 y_1 + 2x_3 x_1 y_6 + 2x_3 x_6 y_1 + x_2^2 y_3 + 2x_2 x_3 y_5 - 2x_4 x_3 y_4$$
$$- x_3^2 y_3 + x_1^2 y_3 - 2x_5 x_2 y_4 + 2x_5 x_3 y_2 + 2x_1 x_6 y_4 - 2x_4 x_2 y_2) e_{24}$$
$$- (2x_1 x_2 y_5 + x_1^2 y_6 - x_3^2 y_6 - 2x_2 x_6 y_2 - 2x_4 x_3 y_1 - x_2^2 y_6 + 2x_5 x_2 y_1$$
$$- 2x_1 x_4 y_4 + 2x_1 x_5 y_2 - 2x_3 x_6 y_4 - 2x_3 x_1 y_3 + 2x_1 x_6 y_1) e_{1234} \in \mathcal{I}.$$

2.10 Cross Ratio of dual unit Quaternions

J3 We have to show that $|e(b)| > |\mathcal{C}(\mathcal{C}\ell_{(3,0,1)})\backslash e(b)|$. Therefore, we search for all elements $d = d_1 e_0 + d_2 e_{1234} \in \mathcal{C}(\mathcal{C}\ell_{(3,0,1)}) \cong \mathbb{D}$ with $\mathfrak{A} - d \in \mathcal{N}(\mathcal{C}\ell_{(3,0,1)})$. If we compute the inverse of $\mathfrak{A} - d$, we get the condition $x_1^2 + x_2^2 + x_3^2 + 2x_1 d_1 + d_1^2 \neq 0$. Since this equation has at most two real solutions at the most, we infer $|\mathcal{C}(\mathcal{C}\ell_{(3,0,1)})\backslash e(b)| = 2$. □

Remark 2.32. *With the identity $\mathcal{C}\ell_{(2,0,1)} = \mathcal{C}\ell_{(2,0,0)} \otimes \mathbb{D}$, see Selig [59, sect. 9.3], we see that the dual unit quaternions define a sphere $S_\mathbb{D}^3$ in four-dimensional dual space.*

A chain defined by three points in $\bigwedge^2 V$ is completely contained in $\bigwedge^2 V$. Therefore, we can interpret the corresponding set of line-symmetric displacements as set of lines. Furthermore, in the quadric model, introduced in section 2.9.1, this is a subspace that corresponds to Study's sphere. Planar sections of Study's sphere with two-dimensional affine submodules result in sets of lines with constant dual angle, i.e., a constant angle and a constant distance with respect to a fixed line. The next step is to show that this is also true in the general case. Therefore, we proof the following:

Theorem 2.20. *A chain $c_{\mathfrak{abc}}$ defined by three arbitrary mutually distant elements $\mathfrak{a}, \mathfrak{b}, \mathfrak{c} \in Spin_{(3,0,1)}$ is characterized by an axis $\mathfrak{l}$ and a dual angle φ_ϵ. All elements $\mathfrak{d}$ contained in $c_{\mathfrak{abc}}$ satisfy $\langle d_\epsilon, l_\epsilon \rangle_\epsilon = \varphi_\epsilon$, where $d_\epsilon, l_\epsilon \in \mathbb{D}^4$ are the dual unit vectors corresponding to $\mathfrak{d}$ and $\mathfrak{l}$.*

Remark 2.33. *The corresponding dual vectors can be obtained directly with the identity $\mathcal{C}\ell_{(2,0,1)} = \mathcal{C}\ell_{(2,0,0)} \otimes \mathbb{D}$ or as the dual number coefficients in the quaternion when using dual unit quaternions.*

Proof. To proof this theorem we use Lemma 2.6. Three mutually distant Spin group elements correspond to Study coordinates of points that admit an unique collineation κ, see Eq. (1.9). This collineation corresponds to a dual quaternion left- or right-multiplication which does not change the cross ratio. We can express this mapping $\gamma \in \mathrm{PGL}(\mathcal{C}\ell_{(3,0,1)}, 2)$ as $\begin{pmatrix} \mathfrak{g} & 0 \\ 0 & 1 \end{pmatrix}$ with $\mathfrak{g} \in Spin_{(3,0,1)}$. The collineation applied to a point in $P = (c, g)^\mathrm{T} \in \mathbb{P}^7(\mathbb{R})$, $c, g \in \mathbb{R}^4$ gives

$$\begin{pmatrix} A & 0 \\ B & A \end{pmatrix} \begin{pmatrix} c \\ g \end{pmatrix} = \begin{pmatrix} Ac \\ Bc + Ag \end{pmatrix}.$$

If we write the elements as dual vectors in $\mathbb{D}^4$, the action can be written as
$$(A + \epsilon B)(c + \epsilon g) = Ac + \epsilon(Ag + Bc).$$
The entries of the dual matrix $\mathcal{O} = A + \epsilon B$ satisfy $\sum_{i=0}^{3} a_i^2 = 1$ and $\sum_{i=0}^{3} a_i b_i = 0$ because the matrix $\mathcal{O}$ corresponds to the action of the Spin group element $\mathfrak{g}$. Furthermore, the matrix $\mathcal{O}$ satisfies $\mathcal{O}\mathcal{O}^T = \mathcal{O}^T\mathcal{O} = I$ and $\det \mathcal{O} = 1$. Thus, $\mathcal{O} \in SO(4, \mathbb{D})$ leaves the dual angle invariant. Chains in $\bigwedge^2 V$ correspond to intersections of two-dimensional affine submodules with $S_{\mathbb{D}}^2$ that correspond to the set of lines with constant dual angle to an axis. If we now apply the inverse mapping $\mathcal{O}^{-1}$ respectively γ^{-1} we get the original chain passing through $\mathfrak{a}, \mathfrak{b}$, and $\mathfrak{c}$. Since the inverse mapping also preserves the dual angle, we have the same property and the theorem is proved. $\square$

Geometrically, we can interpret the set of displacements corresponding to a chain as the set of line-symmetric displacements with respect to a congruence of lines in a reference coordinate system. This line congruence is given by the set of all lines that enclose a constant dual angle with a fixed line.

Remark 2.34. *With remark 2.23 subgroups of Spin groups define subspaces of the chain geometry $\Sigma(\mathcal{C}(\mathcal{C}\ell_{(p,q,r)}), \mathcal{C}\ell_{(p,q,r)})$. The two-parameter group of helical displacements with fixed screw axis is both a subgroup and a $\mathbb{D}$-chain. Therefore, the cross ratio of four elements of a helical displacements group is a dual number.*

For the group $\mathrm{Spin}_{(3,0,1)}$ there is a simple description for the set of all displacements that correspond to elements of a chain. Let us again start, without loss of generality, with a chain of line-symmetric displacements defined by $\mathfrak{a}, \mathfrak{b}, \mathfrak{c} \in \bigwedge^2 V$. That can be achieved with Lemma 2.6. Every axis of each line-symmetric displacement contained in the chain $c_{\mathfrak{abc}}$ is the image of one of the three initial lines under a two-parameter screw group, i.e., all screws with respect to the axes $\mathfrak{l}$ of the cylinder. In fact this group is a one parameter group with a dual parameter. This screw can be represented as the exponential of $\varphi a \in \mathfrak{se}(3)$, where a is the normed axis of the helical motion, $\varphi = \phi + \epsilon d$

2.10 Cross Ratio of dual unit Quaternions

the dual angle and $\mathfrak{se}(3)$ denotes the Lie-algebra of SE(3). Then the chain can be expressed with the sandwich operator

$$c_{\mathfrak{abc}}(\phi, d) = e^{\varphi a} \mathfrak{a} e^{\varphi a *}.$$

For a general chain, i.e., a set of line-symmetric displacements with respect to another reference system, we need the transformation $\mathfrak{g} \in \text{Spin}_{(3,0,1)}$ that corresponds to the collineation κ of Lemma 2.6.

$$c_{\mathfrak{abc}}(\phi, d) = \mathfrak{g} e^{\varphi a} \mathfrak{a} e^{\varphi a *}.$$

Real Cross Ratio Now we ask for a real cross ratio of four dual unit quaternions. First we restrict ourselves again to the case of line-symmetric displacements. For three given line-symmetric displacements $\mathfrak{a}, \mathfrak{b}, \mathfrak{c}$ respectively lines the corresponding $\mathbb{R}$-chain is a closed ruled surface, since $cr(\mathfrak{a}, \mathfrak{b}, \mathfrak{c}, \mathfrak{a}) = 0$, $cr(\mathfrak{a}, \mathfrak{b}, \mathfrak{c}, \mathfrak{b}) = 1$ and $cr(\mathfrak{a}, \mathfrak{b}, \mathfrak{c}, \mathfrak{c}) = \infty$. Every normalized dual vector d_ϵ of the ruled surface satisfies $\langle l_\epsilon, d_\epsilon \rangle_\epsilon = \varphi_\epsilon$, where $\varphi_\epsilon \in \mathbb{D}$ is a constant dual angle and l_ϵ is the axis of the cylinder that defines the congruence, see Th. 2.20. This means that the ruled surface is a closed ruled surface of constant slope. In general its intersection with the ideal plane P_ω is a conic with multiplicity two that has two complex conjugate contact points with multiplicity four with the absolute circle. This can be derived with the help of Plücker coordinates, see section 2.10.2. Hence, the order of the striction curve is reduced by six, see [51, p. 152ff]. The parametrisation of the congruence of lines in terms of Plücker coordinates has degree four. Therefore, the degree of the striction curve is two, i.e., it is a planar curve. In fact, the $\mathbb{R}$-chain defined by $\mathfrak{a}, \mathfrak{b}, \mathfrak{c} \in \bigwedge^2 V$ corresponds to the set of line-symmetric displacements with respect to a closed ruled surface of constant slope whose striction curve is an ellipse, see Fig. 2.4 for example. Note that the striction curve is defined by the planar intersection of the cylinder that is the envelope of the congruence of lines. The plane is defined by the contact points of the three given lines with the cylinder.

2.10.1 Subspaces on Study's Quadric and Sub Chain Geometries

From remark 2.23 we know that subgroups define subchain geometries. Naturally, there are subchain geometries contained in $\mathbb{U}_d$ that can be derived from properties already known from Study's quadric. We examine $\mathbb{R}$-chains, because the set of points of $\mathbb{P}^7(\mathbb{R})$ corresponding to elements of a $\mathbb{D}$-chain spans a four-space in $\mathbb{P}^7(\mathbb{R})$ in general. Therefore, $\mathbb{D}$-chains are not adequate for a comparison with the Study model. We start with generators $\mathcal{G}$, *i.e.*, three-spaces completely contained in Study's quadric. A generator corresponds to the coset of a subgroup if, and only if, the $\dim(\mathcal{G} \cap V^3) \neq 1$, see section 1.2.2. If we use the group $\mathrm{Spin}_{(3,0,1)} \cong \mathbb{U}_d$, we cannot determine chains contained in cosets of SE(2). With elements of the Spin group it is just possible to compute chains that are completely contained in cosets of SO(3), because every element of a chain is distant to every other in cosets of SO(3). If we want to investigate other cosets than cosets of SO(3) we have to use the more general Clifford group, see Eq. (1.31). In this context we can also parametrise chains that corresponds to cosets of SE(2) and even chains contained in subgroups that correspond to lines on Study's quadric. Of course, we lose the property that the whole chain is contained in the Pin group, but we can still normalize after computing the chain, if it is necessary to have Pin or Spin group elements. We demonstrate this method in a simple example.

Example 2.9. *As example we take three Spin group elements that correspond to the same translation group. Let $\mathfrak{a} = e_0 + e_{14}$, $\mathfrak{b} = e_0 + 2e_{14}$, $\mathfrak{c} = e_0 + 4e_{14}$ be elements of $\mathrm{Spin}_{(3,0,1)}$. These elements are not distant to each other, and therefore, there is no chain in the Spin group through these elements. Now we change our setting and take multiples of these element. Thus, let $\bar{\mathfrak{a}} = e_0 + e_{14}$, $\bar{\mathfrak{b}} = 2e_0 + 4e_{14}$, $\bar{\mathfrak{c}} = 4e_0 + 16e_{14}$ be elements of $\Gamma_{(3,0,1)}$. These elements are mutually distant, and therefore, we can compute the $\mathbb{R}$-chain $c_{\bar{\mathfrak{a}}\bar{\mathfrak{b}}\bar{\mathfrak{c}}}(\lambda)$:*

$$c_{\bar{\mathfrak{a}}\bar{\mathfrak{b}}\bar{\mathfrak{c}}}(\lambda) = \frac{2(1+2\lambda)}{2+\lambda}e_0 + \frac{4(2\lambda+1)^2}{(\lambda+2)^2}e_{14},$$

2.10 Cross Ratio of dual unit Quaternions

with $c_{\bar{\mathfrak{a}}\bar{\mathfrak{b}}\bar{\mathfrak{c}}}(0) = \bar{\mathfrak{a}}$, $c_{\bar{\mathfrak{a}}\bar{\mathfrak{b}}\bar{\mathfrak{c}}}(1) = \bar{\mathfrak{b}}$ and $c_{\bar{\mathfrak{a}}\bar{\mathfrak{b}}\bar{\mathfrak{c}}}(\infty) = \bar{\mathfrak{c}}$. To get elements of the Spin group we can normalize and find

$$c_{\bar{\mathfrak{a}}\bar{\mathfrak{b}}\bar{\mathfrak{c}}}(\lambda) = 1 + \frac{2(1+2\lambda)}{2+\lambda}e_{14}.$$

2.10.2 Application to line-symmetric Displacements

In this section we give an example of a chain defined by three line-symmetric displacements $\mathfrak{a}, \mathfrak{b}, \mathfrak{c} \in \bigwedge^2 V$. Let $A = (1, 0, 0, 0, 1, 1)^T \mathbb{R}$, $B = (0, 1, 0, 0, 0, 1)^T \mathbb{R}$, $C = (0, 0, 1, 1, 0, 0)^T \mathbb{R}$ be the axes of three line-symmetric displacements, cf. Ex. 2.8. The corresponding displacements are given by

$$\mathfrak{a} = e_{23} + e_{24} + e_{34}, \qquad \mathfrak{b} = -e_{13} + e_{34}, \qquad \mathfrak{c} = e_{12} + e_{14}.$$

Note, that the minus in front of e_{13} is due to historic reasons, see Ex. 1.6. We compute the chain $c_{\mathfrak{abc}}(\Lambda)$, $\Lambda = \lambda e_0 + \mu e_{1234} \in \mathcal{C}(\mathcal{C}\ell_{(3,0,1)}) \cong \mathbb{D}$, with Eq. (2.8)

$$c_{\mathfrak{abc}}(\Lambda) = \frac{(1-\lambda)e_{23}}{1-\lambda+\lambda^2} - \frac{\lambda e_{13}}{1-\lambda+\lambda^2} + \frac{\lambda(\lambda-1)e_{12}}{1-\lambda+\lambda^2}$$
$$+ \frac{(\lambda^3 - 2\lambda^2 - \mu\lambda + \lambda + 2\mu)\lambda e_{14}}{(1-\lambda+\lambda^2)^2} + \frac{(1 - 2\mu\lambda + \lambda^3 - \lambda + \mu)e_{34}}{(1-\lambda+\lambda^2)^2}$$
$$- \frac{(2\lambda^3 - 4\lambda^2 - \mu\lambda^2 + 3\lambda + \mu - 1)e_{24}}{(1-\lambda+\lambda^2)^2}.$$

In Plücker coordinates the line congruence can be written as

$$l(\lambda, \mu) = \begin{pmatrix} (1-\lambda+\lambda^2)(1-\lambda) \\ (1-\lambda+\lambda^2)\lambda \\ (1-\lambda+\lambda^2)\lambda(\lambda-1) \\ (\lambda^3 - 2\lambda^2 - \mu\lambda + \lambda + 2\mu)\lambda \\ -2\lambda^3 + 4\lambda^2 + \mu\lambda^2 - 3\lambda - \mu + 1 \\ 1 - 2\mu\lambda + \lambda^3 - \lambda + \mu \end{pmatrix}.$$

Intersection with the absolute Circle To determine the intersection of the congruence of lines with the absolute circle, we intersect its ideal curve given by

$$c(\lambda) = \left((1-\lambda+\lambda^2)(1-\lambda),\ (1-\lambda+\lambda^2)\lambda,\ (1-\lambda+\lambda^2)\lambda(\lambda-1)\right)^T$$

with the absolute circle determined by $x_1^2 + x_2^2 + x_3^2 = 0$. This leads to the equation

$$1 - 4\lambda + 10\lambda^2 - 16\lambda^3 + 19\lambda^4 - 16\lambda^5 + 10\lambda^6 - 4\lambda^7 + \lambda^8 = 0$$

with solutions

$$\lambda_1 = \lambda_2 = \lambda_3 = \lambda_4 = \frac{1}{2} - \frac{i}{2}\sqrt{3}, \ \lambda_5 = \lambda_6 = \lambda_7 = \lambda_8 = \frac{1}{2} + \frac{i}{2}\sqrt{3}.$$

Hence, this congruence of lines has two points in common with the absolute circle. At these points it has contact of order three.

Axis of the Congruence and constant dual Angle We are interested in the axis of this congruence, *i.e.*, in the dual vector $a_\epsilon \in \mathbb{D}^3$ with

$$\langle a_\epsilon, c_{\mathfrak{abc}}(\Lambda) \rangle_\epsilon = k_1 + \epsilon k_2, \ \text{for all } \Lambda \in \mathbb{D}$$

where $c_{\mathfrak{abc}}(\Lambda)$ is the chain defined $\mathfrak{a}, \mathfrak{b}, \mathfrak{c}$, and $\Lambda = \lambda + \epsilon\mu$ is the cross ratio, and $k_1, k_2 \in \mathbb{R}$ are constants. Therefore, we look for the axis $A = (c, \bar{c}) = (c_1, c_2, c_3, c_4, c_5, c_6)^\mathrm{T}$ of the linear complex that contains $l(\lambda, \mu)$. This is done with the approach

$$\Omega(A, l(\lambda, \mu)) = c_1 l_4(\lambda, \mu) + c_2 l_5(\lambda, \mu) + c_3 l_6(\lambda, \mu)$$
$$+ c_4 l_1(\lambda, \mu) + c_5 l_2(\lambda, \mu) + c_6 l_3(\lambda, \mu) \stackrel{!}{=} 0. \qquad (2.17)$$

For a more detailed description of this implicitization method, see [63]. After expanding Eq. (2.17) and collecting the coefficients of $\lambda^i \mu^j$, $i = 0, \ldots, 4$, $j = 0, 1$ we get

$$\Omega(A, l(\lambda, \mu)) = (c_6 + c_1)\lambda^4 + (-2c_1 - 2c_2 + c_3 - c_4 + c_5 - 2c_6)\lambda^3$$
$$+ (-c_1 + c_2)\lambda\mu + (+c_1 + 4c_2 + 2c_4 - c_5 + 2c_6)\lambda^2$$
$$+ (2c_1 - 2c_3)\lambda\mu + (-3c_2 - c_3 - 2c_4 + c_5 - c_6)\lambda$$
$$+ (-c_2 + c_3)\mu + (c2 + c3 + c4) \stackrel{!}{=} 0.$$

In order to find a solution, every coefficient written in brackets has to vanish. We can reformulate this as matrix vector product:

2.10 Cross Ratio of dual unit Quaternions

$$M \cdot A = \begin{pmatrix} 1 & 0 & 0 & 0 & 0 & 1 \\ -2 & -2 & 1 & -1 & 1 & -2 \\ -1 & 1 & 0 & 0 & 0 & 0 \\ 1 & 4 & 0 & 2 & -1 & 2 \\ 2 & 0 & -2 & 0 & 0 & 0 \\ 0 & -3 & -1 & -2 & 1 & -1 \\ 0 & -1 & 1 & 0 & 0 & 0 \\ 0 & 1 & 1 & 1 & 0 & 0 \end{pmatrix} \cdot \begin{pmatrix} c_1 \\ c_2 \\ c_3 \\ c_4 \\ c_5 \\ c_6 \end{pmatrix} = \begin{pmatrix} 0 \\ 0 \\ 0 \\ 0 \\ 0 \\ 0 \\ 0 \\ 0 \end{pmatrix}.$$

The kernel of M is determined by

$$\ker M = (\nu, \nu, \nu, -2\nu, -\nu, -\nu)^T, \; \nu \in \mathbb{R}.$$

We see that the kernel of M is one-dimensional, geometrically this means that $l(\lambda, \mu)$ is contained in a hyperplane, i.e., a four-space. In a next step, we could assume that there is a quadratic complex of lines that contains the lines corresponding to elements of $l(\lambda, \mu)$ and evaluate a quadratic approach to look for quadrics in $\mathbb{P}^5(\mathbb{R})$ that contain $l(\lambda, \mu)$. We know that the line congruence $l(\lambda, \mu)$ is contained in a linear complex, cf. Def. 1.6. If the linear complex is singular then the Plücker condition (1.12) is satisfied. If $\Omega(A, A) = 0$ the point A corresponds to a line and the complex is singular, i.e., it is the set of all lines that meet the line represented by $(c, \bar{c})$. If $\Omega(A, A) \neq 0$, the linear complex is regular and it consists of all path normals of a helical motion. For a regular linear complex we can calculate the pitch p and the axis $(a, \bar{a})$ with Eq. (1.14). For our example this results in

$$p = \frac{c \cdot \bar{c}}{c^2} = -\frac{4}{3}, \quad (a, \bar{a}) = (c, \bar{c} - pc) = \left((-1, -1, -1)^T, (\frac{2}{3}, -\frac{1}{3}, -\frac{1}{3})^T\right)$$
$$\hat{=} ((3, 3, 3)^T, (-2, 1, 1)^T).$$

Remark 2.35. *In Klein's model this hyperplane has a two-dimensional intersection with the image-plane of the field of ideal lines V^2 given by $x_0 = x_1 = x_2 = 0$. Due to the fact, that we can map every chain to any other chain and the zero divisor is invariant under multiplication with an arbitrary element, each chain is contained in a four-space that intersects the exceptional generator in Study's model with dimension two.*

To compute the dual angle we switch the model to $\mathcal{C}\ell_{(3,0,0)}(\mathbb{D})$ and rewrite the line congruence (see Eq. (2.15)) as:

$$c_{abc}(\lambda, \mu) = \left(\frac{1-\lambda}{\lambda^2 - \lambda + 1} + \epsilon \frac{\lambda(\lambda^3 - 2\lambda^2 + \mu\lambda + \lambda - 2\mu)}{(\lambda^2 - \lambda + 1)^2} \right) e_1$$
$$+ \left(\frac{\lambda}{\lambda^2 - \lambda + 1} + \epsilon \frac{(1-\lambda)(2\lambda^2 - 2\lambda + \mu\lambda + 1 + \mu)}{(\lambda^2 - \lambda + 1)^2} \right) e_2$$
$$+ \left(\frac{(\lambda - 1)\lambda}{\lambda^2 - \lambda + 1} + \epsilon \frac{\lambda^3 - \lambda + 1 - \mu + 2\mu\lambda}{(\lambda^2 - \lambda + 1)^2} \right) e_3.$$

This is done because the dual scalar product can be expressed in $\mathcal{C}\ell_{(3,0,0)}(\mathbb{D})$ with help of the geometric product. The axis $A = (a, \bar{a})^T \mathbb{R}$ is transferred to this model and normalized:

$$\mathfrak{A} = \frac{1}{9}\sqrt{3}((3 - 2\epsilon)e_1 + (3 + \epsilon)e_2 + (3 + \epsilon)e_3).$$

The dual scalar product can be calculated with Eq. (1.27)

$$\langle c_{abc}(\lambda, \mu), a_\epsilon \rangle_\epsilon = c_{abc}(\lambda, \mu) \cdot \mathfrak{A} = \frac{1}{3}\sqrt{3} + \frac{4}{9}\sqrt{3}\epsilon,$$

where a_ϵ denotes the dual unit vector corresponding to $\mathfrak{A}$.

Ruled Surface of constant Slope Now we give an example of an $\mathbb{R}$-chain. Therefore, we have to assume a real cross ratio in Eq. (2.8). In our example it is sufficient to set $\mu = 0$. This leads to a curve on M_2^4 in Plücker coordinates:

$$l(\lambda) = \begin{pmatrix} (1 - \lambda + \lambda^2)(1 - \lambda) \\ (1 - \lambda + \lambda^2)\lambda \\ (1 - \lambda + \lambda^2)\lambda(\lambda - 2) \\ (\lambda^3 - 2\lambda^2 + \lambda)\lambda \\ -2\lambda^3 + 4\lambda^2 - 3\lambda + 1 \\ 1 + \lambda^3 - \lambda \end{pmatrix}.$$

The striction curve of every $\mathbb{R}$-chain in $\bigwedge^2 V$ is an ellipse that may degenerate into a line or a point. As example we calculate the striction curve of the ruled surface given above. Therefore, we describe it in Euclidean space. A point p on the line given by its Plücker coordinates

2.10 Cross Ratio of dual unit Quaternions

$(l, \bar{l})$ can be calculated by $p = \frac{l \times \bar{l}}{\sqrt{l \cdot l}}$. The vector $r = \frac{l}{\sqrt{l \cdot l}}$ is the direction vector of the generator. Therefore, we parametrise the ruled surface by using a curve on the surface together with the direction of each generator with

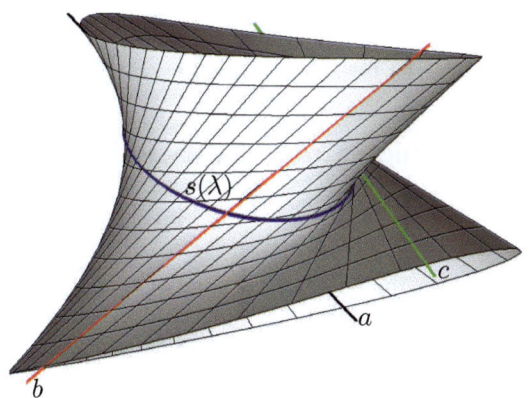

Figure 2.4: Ruled surface of constant slope with striction curve

$$f(\lambda, t) = \frac{1}{(\lambda^2 - \lambda + 1)^2} \left[\begin{pmatrix} 2\lambda^3 - 3\lambda^2 + 2\lambda \\ \lambda^4 - 2\lambda^3 + \lambda^2 + \lambda - 1 \\ -\lambda^3 + 3\lambda^2 - 3\lambda + 1 \end{pmatrix} + t \begin{pmatrix} -\lambda^3 + 2\lambda^2 - 2\lambda + 1 \\ \lambda^3 - \lambda^2 + \lambda \\ \lambda^4 - 2\lambda^3 + 2\lambda^2 - \lambda \end{pmatrix} \right].$$

The striction curve of this ruled surface results in

$$s(\lambda) = \frac{1}{\lambda^2 - \lambda + 1} \left(\lambda, \ \lambda^2 - 1, \ \lambda^2 - 2\lambda + 1 \right)^{\mathrm{T}}.$$

This is a rational parametrisation of an ellipse. A visualization of the ruled surface of constant slope together with its striction curve is presented in Fig. 2.4.

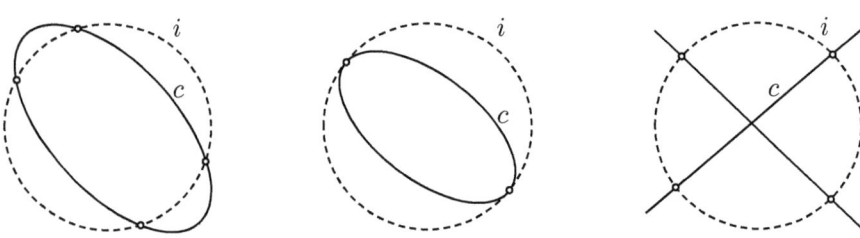

Figure 2.5: Intersection of the ideal curves with the absolute circle.

2.10.3 Dual Quaternion Cross Ratio and Conics on Study's Quadric

With the interpretation of the cross ratio of dual unit quaternions, we take a look at conic sections of Study's quadric. In this section we answer the following question: Let $\bar{A}, \bar{B}, \bar{C} \in S_2^6$ be three points contained in Study's quadric that span the two-space P_1^2. Is there a point $\bar{D} \in P_1^2 \cap S_2^6$ with $cr(\mathfrak{a}, \mathfrak{b}, \mathfrak{c}, \mathfrak{d}) \in \mathbb{D}$, where $\mathfrak{a}, \mathfrak{b}, \mathfrak{c}, \mathfrak{d}$ are the corresponding Spin group elements or dual unit quaternions. As we know from section 2.10 we can always map two-spaces that intersect S_2^6 to two-spaces whose intersection with S_2^6 can be interpreted as set of line-symmetric displacements. Therefore, it is sufficient to study this question for line-symmetric displacements. Thus, we restrict ourselves to Klein's quadric. A conic on Klein's quadric corresponds to a regulus, see Def. 1.5. Hence, the question can be formulated as follows:

Q: Is there an additional line except the three given lines that belongs to the regulus $\mathfrak{R}$ and to the line congruence $\mathfrak{L}$ given by the chain?

To answer this question we examine the intersections of $\mathfrak{R}$ and $\mathfrak{L}$ interpreted as set of points in $\mathbb{P}^3(\mathbb{R})$ with the ideal plane P_ω given by $x_0 = 0$. Furthermore, we denote the ideal points contained by the lines corresponding to $\mathfrak{a}, \mathfrak{b}, \mathfrak{c}$ by A, B, C. To classify the conic section c of $\mathfrak{R}$ with the ideal plane we intersect it with the absolute circle given by $i: x_1^2 + x_2^2 + x_3^2 = 0$. This leads to three different cases corresponding to a general hyperboloid, see Fig. 2.5 (left), a hyperboloid of revolution,

2.10 Cross Ratio of dual unit Quaternions

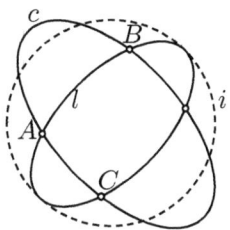 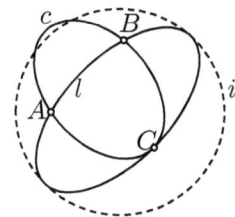 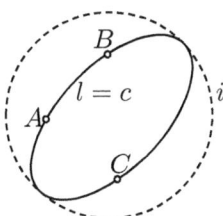

Figure 2.6: Intersection of conic sections in P_ω

Fig. 2.5 (middle) and a hyperbolic paraboloid or a degenerate conic, i.e., two pencils of lines with one line in common, Fig. 2.5 (right). The congruence of lines $\mathfrak{L}$ contains a hyperboloid of revolution and all other lines are parallel to the lines of this hyperboloid. Hence, the intersection of $\mathfrak{L}$ with P_ω results in the same figure as for the hyperboloid of revolution, see Fig. 2.6 (right). To answer **Q** we have to intersect the conics in the ideal plane. Since two quadratic curves have in general four, possibly complex, points of intersection and we already know three of these four points, we can assume that the fourth point of intersection is real. Therefore, there are three different possible cases, see Fig. 2.6. For hyperbolic paraboloids the intersection of the congruence of lines with the ideal plane and the intersection of the ruled surface itself with the ideal plane are identical in every case. The results we achieve for hyperboloids also hold for hyperbolic paraboloids. All in all the following cases can occur: There are three intersection points and one point is a double point, there are four different points, or the conics are identical. Now we formulate this as a theorem.

Theorem 2.21. *Let $A, B, C \in S_2^6 \subseteq \mathbb{P}^7(\mathbb{R})$ be three points that are not joined by a line contained in S_2^6, i.e., the conic $c(t_1 : t_0)$ defined by A, B, C is regular. Furthermore, let $\mathfrak{a}, \mathfrak{b}, \mathfrak{c} \in Spin_{(3,0,1)}$ be the corresponding Spin group elements. Then, one of the three statements is true:*

(1) There is no point $D \in c(t_1 : t_0)$ with $cr(\mathfrak{a}, \mathfrak{b}, \mathfrak{c}, \mathfrak{d}) \in \mathbb{D}$.

(2) There is exactly one point $D \in c(t_1 : t_0)$ with $cr(\mathfrak{a}, \mathfrak{b}, \mathfrak{c}, \mathfrak{d}) \in \mathbb{D}$.

(3) For every point $D \in c(t_1:t_0)$ the cross ratio $cr(\mathfrak{a}, \mathfrak{b}, \mathfrak{c}, \mathfrak{d})$ is a dual number.

To show that each case occurs we give examples for every case.

Example 2.10. *We take the chain from Ex. 2.8 given by Eq. (2.15). This chain is generated by*

$$\mathfrak{a} = e_1 + \epsilon e_2 + \epsilon e_3, \quad \mathfrak{b} = e_2 + \epsilon e_3, \quad \mathfrak{c} = \epsilon e_1 + e_3,$$

with corresponding Plücker coordinates given by $A = (1,0,0,0,1,1)^T \mathbb{R}$, $B = (0,1,0,0,0,1)^T \mathbb{R}$ and $C = (0,0,1,1,0,0)^T \mathbb{R}$. Now we parametrise the conic on M_2^4 defined by these three points with a homogeneous parameter $(t_1:t_0)$.

$$c(t_1:t_0) = t_1 t_0 \Omega(A,B)C + (t_0^2 - t_1 t_0)\Omega(A,C)B + (t_1^2 - t_1 t_0)\Omega(C,B)A$$
$$= (t_1^2 - t_0 t_1, 2t_0^2 - 2t_1 t_0, t_1 t_0, t_1 t_0, t_1^2 - t_1 t_0, t_1^2 - t_1 t_0 + t_0^2)^T.$$

The chain from Ex. 2.8 is given by

$$c_{\mathfrak{abc}}(\lambda, \mu) = \left(\frac{1-\lambda}{\lambda^2 - \lambda + 1} + \epsilon \frac{\lambda(\lambda^3 - 2\lambda^2 + \mu\lambda + \lambda - 2\mu)}{(\lambda^2 - \lambda + 1)^2} \right) e_1$$
$$+ \left(\frac{\lambda}{\lambda^2 - \lambda + 1} + \epsilon \frac{(1-\lambda)(2\lambda^2 - 2\lambda + \mu\lambda + 1 + \mu)}{(\lambda^2 - \lambda + 1)^2} \right) e_2$$
$$+ \left(\frac{(\lambda - 1)\lambda}{\lambda^2 - \lambda + 1} + \epsilon \frac{\lambda^3 - \lambda + 1 - \mu + 2\mu\lambda}{(\lambda^2 - \lambda + 1)^2} \right) e_3.$$

If we compare this with the results of section 2.10.2, we see that the resulting chain is exactly the same. The intersection of $c(t_1:t_0)$ with P_ω is given by

$$c_1(t_1:t_0) = (t_1^2 - t_0 t_1, 2t_0^2 - 2t_1 t_0, t_1 t_0)^T.$$

Moreover, the intersection of $c_{\mathfrak{abc}}(\lambda, \mu)$ with P_ω can be represented in homogeneous coordinates by

$$c_2(t_1:t_0) = (t_0^2 - t_0 t_1, t_0 t_1, t_1^2 - t_0 t_1)^T.$$

To intersect c_1 and c_2 we need the implicit equation of one of them. Therefore, we look for a collineation in $\mathbb{P}^2(\mathbb{R})$ that transforms the

2.10 Cross Ratio of dual unit Quaternions

standard conic parametrised by $p(t_1:t_0) = (t_0^2, t_0t_1, t_1^2)^T$ with implicit form $y_0y_2 - y_1^2 = 0$ to $c_2(t_1:t_0)$. This can be realized by

$$\begin{pmatrix} 1 & -1 & 0 \\ 0 & 1 & 0 \\ 0 & -1 & 1 \end{pmatrix} \begin{pmatrix} t_0^2 \\ t_0t_1 \\ t_1^2 \end{pmatrix} = \begin{pmatrix} t_0^2 - t_0t_1 \\ t_0t_1 \\ t_1^2 - t_0t_1 \end{pmatrix} = \begin{pmatrix} x_0 \\ x_1 \\ x_2 \end{pmatrix}.$$

The inverse collineation maps the conic to the standard conic

$$\begin{pmatrix} 1 & 1 & 0 \\ 0 & 1 & 0 \\ 0 & 1 & 1 \end{pmatrix} \begin{pmatrix} x_0 \\ x_1 \\ x_2 \end{pmatrix} = \begin{pmatrix} x_0 + x_1 \\ x_1 \\ x_1 + x_2 \end{pmatrix} = \begin{pmatrix} y_0 \\ y_1 \\ y_2 \end{pmatrix}.$$

With the implicit form of the standard conic we can now determine the implicit form of $c_2(t_1:t_0)$ as

$$y_0y_2 - y_1^2 = (x_0 + x_1)(x_1 + x_2) - x_1^2 = x_0x_1 + x_0x_2 + x_1x_2 = 0.$$

Now we are able to intersect both conics, i.e., we insert the parameter form of c_1 in the implicit form of c_2. This leads to the equation

$$t_0^2 t_1^2 - t_0 t_1^3 = 0.$$

There are four solutions given by the homogeneous parameters

$$(t_1:t_0)_1 = (0:1),\ (t_1:t_0)_2 = (0:1),\ (t_1:t_0)_3 = (1:1),\ (t_1:t_0)_4 = (1:0).$$

We see that solution 1 and solution 2 are identical. Thus, there is no point $D \in c(t_1:t_2)$ with corresponding algebra element $\mathfrak{d}$ satisfying $cr(\mathfrak{a}, \mathfrak{b}, \mathfrak{c}, \mathfrak{d}) \in \mathbb{D}$, and we have case (1) of Th. 2.21. Now we do the same calculation but we start with $\mathfrak{a}, \mathfrak{b}$, and

$$\mathfrak{c} = c_{\mathfrak{abc}}(2,0) = \left(-\frac{1}{3} + \frac{4}{9}\epsilon\right)e_1 + \left(\frac{2}{3} - \frac{5}{9}\epsilon\right)e_2 + \left(\frac{2}{3} + \frac{7}{9}\epsilon\right)e_3.$$

The corresponding Plücker coordinate vector is given by

$$C = (-3, 6, 6, 4, -5, 7)^T \mathbb{R}.$$

Of course, this results in the same chain, and therefore, in the same ideal conic of the congruence of lines. The regulus defined by the points A, B, C can be parametrised by

$$\bar{c}(t_1:t_0) = (-4t_0t_1 + t_1^2, -10t_1t_0 + 16t_0^2, 6t_1t_0,$$

$$4t_0t_1, -6t_1t_0 + t_1^2, 16t_0^2 - 10t_1t_0 + t_1^2)^T\mathbb{R}.$$

Its intersection with the ideal plane P_ω of $\mathbb{P}^3(\mathbb{R})$ is

$$c_3(t_1:t_0) = (-t_1(4t_0 - t_1), 2t_0(-5t_1 + 8t_0), 6t_1t_0)^T.$$

The intersection of c_2 with c_3 leads to

$$7t_1^2t_0^2 + 8t_1t_0^3 - t_1^3t_0 = 0.$$

This yields

$$(t_1:t_0)_1 = (0:1), \quad (t_1:t_0)_2 = (1:0), \quad (t_1:t_0)_3 = (8:-1), \quad (t_1:t_0)_4 = (1:1).$$

We see that there is a solution that does not correspond to the start points A, B, or C. Therefore, we have

$$D = \bar{c}(8:-1) = 16(6, 6, -3, -2, 7, 10)^T \mathbb{R}.$$

As dual vector this line has the form

$$\mathfrak{d} = \frac{1}{16}((6 - 2\epsilon)e_1 + (6 + 7\epsilon)e_2 + (-3 + 10\epsilon)e_3)$$

and after normalization the corresponding Spin group element is given by

$$\mathfrak{d} = \frac{1}{9}((6 - 2\epsilon)e_1 + (6 + 7\epsilon)e_2 + (-3 + 10\epsilon)e_3).$$

Finally, we verify that the cross ratio is a dual number

$$cr(\mathfrak{a}, \mathfrak{b}, \mathfrak{c}, \mathfrak{d}) = (\mathfrak{d} - \mathfrak{a})(\mathfrak{b} - \mathfrak{a})^{-1}(\mathfrak{b} - \mathfrak{c})(\mathfrak{d} - \mathfrak{c})^{-1} = \frac{1}{3} + \epsilon\frac{2}{9}.$$

Hence, we arrive at case (2) of Th. 2.21.

In the third case, every point on the conic leads to a dual cross ratio. This situation occurs if the regulus is a hyperboloid of revolution. In fact, the cross ratio is a real number in this case.

Remark 2.36. *For the hyperbolic paraboloid we have the same cases. All lines define a dual cross ratio with three given lines if all lines of one regulus intersect the lines of the other regulus orthogonally. If this is not the case, there may be one line with dual cross ratio or there may be no one.*

2.11 Chains of Geometric Entities

Remark 2.37. *For general conics on S_2^6 there are two possibilities to perform this calculation. The first one is based on the transformation presented in section 2.10. After the transformation is performed we can proceed with line-symmetric displacements. The second method is to take the conic that is defined by the first four Study coordinates and intersect it with the conic defined by the first four Study coordinates corresponding to the chain that is defined by the three points.*

2.11 Chains of Geometric Entities

Chains can be realized in every Clifford algebra. The corresponding geometric entities differ. Thus, it is interesting to examine the properties of the chain with respect to the geometric meaning. Of course, we can not do this in detail. Furthermore, it can be helpful to use subgroups to obtain specialized chains. This means that a chain in $\bigwedge^1 V$ of QnGA defined by three general inversions in hyperquadrics in principal position except for translation, delivers a chain where every element is again an inversion in a hyperquadric in principal position. If we take three inversions in hyperspheres, *i.e.*, three elements of a subgroup, the entire chain consists of inversions in hyperspheres, see remark 2.23. Furthermore, we can interpret these inversions as the corresponding geometric entities.

2.11.1 Quadric Geometric Algebra

In section 1.10 we introduced the quadric geometric algebra. In this algebra we can perform inversions with respect to hyperquadrics in principal position. In this section we give two examples for the three-dimensional case to show properties of chains we have examined in the previous sections. The first example is the chain $c_{\mathfrak{abc}}$ that is defined by three algebra elements. These three elements were generated with the help of the map $\eta : P \in \mathbb{R}^3 \mapsto \bigwedge^1 V(\mathcal{C}\ell_{(6,3,0)})$, see Eq. (1.59).

$$\begin{aligned}
\mathfrak{a} = &\left(\eta(-1,0,0)^{\mathrm{T}} \wedge \eta(1,0,0)^{\mathrm{T}} \wedge \eta(0,1,0)^{\mathrm{T}} \right. \\
&\left. \wedge \eta(0,-1,0)^{\mathrm{T}} \wedge \eta(0,0,1)^{\mathrm{T}} \wedge \eta(0,0,-1)^{\mathrm{T}}\right) \cdot \mathfrak{J} \\
= &-6e_1 + e_3 - 6e_4 + e_6 - 6e_7 + e_9,
\end{aligned}$$

$$\mathfrak{b} = \left(\eta(\tfrac{7}{2},0,0)^{\mathrm{T}} \wedge \eta(\tfrac{13}{2},0,0)^{\mathrm{T}} \wedge \eta(5,\tfrac{3}{2},0)^{\mathrm{T}}\right.$$
$$\left.\wedge\, \eta(5,-\tfrac{3}{2},0)^{\mathrm{T}} \wedge \eta(5,0,\tfrac{3}{2})^{\mathrm{T}} \wedge \eta(5,0,-\tfrac{3}{2})^{\mathrm{T}}\right)\cdot\mathfrak{J}$$
$$= -\frac{6561}{64}e_1 - \frac{32805}{64}e_2 - \frac{199017}{512}e_3 - \frac{6561}{64}e_4$$
$$- \frac{199017}{512}e_6 - \frac{6561}{64}e_7 - \frac{199017}{512}e_9,$$
$$\mathfrak{c} = \left(\eta(4,4,0)^{\mathrm{T}} \wedge \eta(6,4,0)^{\mathrm{T}} \wedge \eta(5,5,0)^{\mathrm{T}}\right.$$
$$\left.\wedge\, \eta(5,3,0)^{\mathrm{T}} \wedge \eta(5,4,1)^{\mathrm{T}} \wedge \eta(5,4,-1)^{\mathrm{T}}\right)\cdot\mathfrak{J}$$
$$= -6e_1 - 30e_2 - 40e_3 - 6e_4 - 24e_5 - 40e_6 - 6e_7 - 40e_9.$$

After normalization we get

$$\mathfrak{a} = -e_1 + \tfrac{1}{6}e_3 - e_4 + \tfrac{1}{6}e_6 - e_7 + \tfrac{1}{6}e_9,$$
$$\mathfrak{b} = -\tfrac{2}{3}e_1 - \tfrac{10}{3}e_2 - \tfrac{91}{36}e_3 - \tfrac{2}{3}e_4 - \tfrac{91}{36}e_6 - \tfrac{2}{3}e_7 - \tfrac{91}{36}e_9,$$
$$\mathfrak{c} = -e_1 - 5e_2 - \tfrac{20}{3}e_3 - e_4 - 4e_5 - \tfrac{20}{3}e_6 - e_7 - \tfrac{20}{3}e_9.$$

Note, that the normalization does not change the GIPNS of $\mathfrak{a}, \mathfrak{b}$, or $\mathfrak{c}$. The geometric inner product null spaces are given by

$$\mathbb{NI}_G(\mathfrak{a}) = \{(x,y,z)^{\mathrm{T}} \in \mathbb{R}^3 \mid x^2+y^2+z^2-1=0\},$$
$$\mathbb{NI}_G(\mathfrak{b}) = \{(x,y,z)^{\mathrm{T}} \in \mathbb{R}^3 \mid x^2+y^2+z^2-10x+\tfrac{91}{4}=0\},$$
$$\mathbb{NI}_G(\mathfrak{c}) = \{(x,y,z)^{\mathrm{T}} \in \mathbb{R}^3 \mid x^2+y^2+z^2-10x-8y+40=0\}.$$

The GIPNS of $\mathfrak{a}$ is the unit sphere centered at the origin, the quadric corresponding to $\mathfrak{b}$ is the sphere centered at $M = (5,0,0)^{\mathrm{T}}$ with radius $r = \tfrac{3}{2}$ and $\mathfrak{c}$ is an inner product sphere with radius $r=1$ and center $M = (5,4,0)^{\mathrm{T}}$, see Fig. 2.7 (left). We compute the chain $c_{\mathfrak{abc}}$ defined by the three elements:

$$c_{\mathfrak{abc}}(\lambda) = -\frac{1}{3}\frac{(63+2\lambda+99\lambda^2)e_1}{28\lambda+33\lambda^2+21} - \frac{5}{3}\frac{\lambda(99\lambda+65)e_2}{28\lambda+33\lambda^2+21}$$
$$- \frac{1}{18}\frac{(-63-166\lambda+3960\lambda^2)e_3}{28\lambda+33\lambda^2+21} - \frac{1}{3}\frac{(63+2\lambda+99\lambda^2)e_4}{(28\lambda+33\lambda^2+21)}$$

2.11 Chains of Geometric Entities

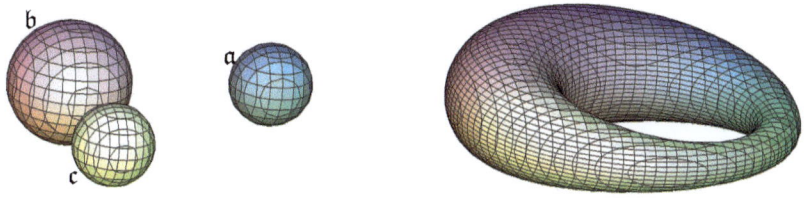

Figure 2.7: Chain in quadric geometric algebra

$$-\frac{132\lambda(\lambda-1)e_5}{28\lambda+33\lambda^2+21} - \frac{1}{18}\frac{(-63-166\lambda+3960\lambda^2)e_6}{(28\lambda+33\lambda^2+21)}$$
$$-\frac{1}{3}\frac{(63+2\lambda+99\lambda^2)e_7}{28\lambda+33\lambda^2+21} - \frac{1}{18}\frac{(-63-166\lambda+3960\lambda^2)e_9}{28\lambda+33\lambda^2+21}.$$

From Eq. (1.55) we see, that the geometric inner product null space of a vector determines a sphere if the coefficients of e_1, e_4 and e_7 are equal. Thus, every entity contained in c_{abc} corresponds to a sphere, see Fig. 2.7 (right) for the cyclide that is the envelope of this set of spheres. The GIPNS of the chain $c_{abc}(\lambda)$ can be determined as

$$\mathbb{NI}_G(c_{abc}(\lambda)) = \{(x,y,z)^T \in \mathbb{R}^3 \mid -990\lambda^2 x - 650\lambda x + 63x^2 + 2\lambda x^2 + 99\lambda^2 x^2$$
$$- 63 - 166\lambda + 3960\lambda^2 + 63y^2 + 2\lambda y^2 + 99\lambda^2 y^2 - 792\lambda^2 y$$
$$+ 792\lambda y + 63z^2 + 2\lambda z^2 + 99\lambda^2 z^2 = 0, \quad \forall \lambda \in \mathbb{R}\}.$$

The implicit equation of the envelope is determined by

$$6236x^4 - 61720x^3 - 50688x^2 y + 12472x^2 y^2 + 137784x^2 + 12472x^2 z^2$$
$$-61720xy^2 + 257400xy + 8420x - 61720xz^2 - 256369 + 243409z^2 + 6236y^4$$
$$-50688yz^2 + 12472y^2 z^2 + 115632y + 6236z^4 - 50688y^3 + 86593y^2 = 0.$$

Furthermore, the chain is defined by three grade-1 elements corresponding to inner product spheres that can also be interpreted as Pin group elements or as inversions in these spheres. Since the group of conformal transformations is a subgroup of the group of reflections in quadrics in principal position it also defines a sub chain geometry, see remark 2.23.

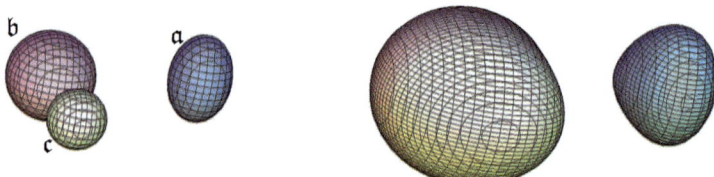

Figure 2.8: Chain in quadric geometric algebra

Remark 2.38. *Algebra elements corresponding to inner product spheres with identical radii form a subchain geometry. Thus, a chain generated by three spheres with radius $r = c_0 \in \mathbb{R}$ contains only spheres with radius c_0. Furthermore, the centers of the spheres are contained in a circle.*

The second example we examine is given by the three quadrics that are the GIPNS of

$$\mathfrak{a} = -36e_1 + 6e_3 - 36e_4 + 6e_6 - 18e_7 - 9e_8 + 6e_9,$$
$$\mathfrak{b} = -24e_1 - 120e_2 - 91e_3 - 24e_4 - 91e_6 - 24e_7 - 91e_9,$$
$$\mathfrak{c} = -3e_1 - 15e_2 - 20e_3 - 3e_4 - 12e_5 - 20e_6 - 3e_7 - 20e_9.$$

The spheres corresponding to $\mathfrak{b}$ and $\mathfrak{c}$ are identical with the spheres of the previous example. Only the quadric corresponding to $\mathfrak{a}$ was changed, it is an ellipsoid, see Fig. 2.8 (left) given by

$$2x^2 + 2y^2 + z^2 - z - 2 = 0.$$

Furthermore, we did not normalize the elements since this does not change the GIPNS and a calculation without normalization is much faster. The chain $c_{\mathfrak{abc}}$ results in

$$c_{\mathfrak{abc}}(\lambda) = -\frac{3(1997\lambda^2 - 817\lambda + 2964)e_1}{-1726\lambda + 1997\lambda^2 + 247} - \frac{15\lambda(1997\lambda + 2147)e_2}{-1726\lambda + 1997\lambda^2 + 247}$$
$$- \frac{2(-741 + 4340\lambda + 19970\lambda^2)e_3}{-1726\lambda + 1997\lambda^2 + 247} - \frac{3(1997\lambda^2 - 817\lambda + 2964)e_4}{-1726\lambda + 1997\lambda^2 + 247}$$
$$- \frac{23964\lambda(\lambda - 1)e_5}{-1726\lambda + 1997\lambda^2 + 247} - \frac{2(-741 + 4340\lambda + 19970\lambda^2)e_6}{-1726\lambda + 1997\lambda^2 + 247}$$

$$-\frac{3(665\lambda + 1482 + 1997\lambda^2)e_7}{-1726\lambda + 1997\lambda^2 + 247} + \frac{2223(\lambda - 1)e_8}{-1726\lambda + 1997\lambda^2 + 247}$$
$$-\frac{2(-741 + 4340\lambda + 19970\lambda^2)e_9}{-1726\lambda + 1997\lambda^2 + 247}.$$

Again we compute the GIPNS of the chain $c_{\mathfrak{abc}}$ and get

$$\begin{aligned}\mathbb{NI}_G(c_{\mathfrak{abc}}(\lambda)) = \{(x,y,z)^\mathrm{T} \in \mathbb{R}^3 \mid & 52080\lambda - 4446z - 2451\lambda x^2 + 4446z^2 + 4446\lambda z \\ & - 2451\lambda y^2 - 8892 - 64410\lambda x + 1995\lambda z^2 + 47928\lambda y + 5991\lambda^2 x^2 \\ & + 239640\lambda^2 + 5991\lambda^2 z^2 - 47928\lambda^2 y - 59910\lambda^2 x + 5991\lambda^2 y^2 \\ & + 8892 x^2 + 8892 y^2 = 0, \, \forall \lambda \in \mathbb{R}\}.\end{aligned}$$

As parametrisation of the family of quadrics in principal position. We determine the envelope of this family, see Fig. 2.8 (right)

$$\begin{aligned}&-3745280640 + 2946608160x - 1095825792y - 1574951040z + 1472358564x^2 \\ &+ 2089543536y^2 + 1273701252z^2 + 2058028320xy + 109803774y^2z^2 \\ &- 347861424z^2y + 546057720zx + 142058592zy - 28249884zx^2 \\ &- 28249884zy^2 + 138053658x^2y^2 + 109803774x^2z^2 - 489920016x^2y \\ &- 815538900xy^2 - 269481180xz^2 + 69026829y^4 - 489920016y^3 + 34187973z^4 \\ &- 41427828z^3 + 69026829x^4 - 815538900x^3 = 0.\end{aligned}$$

Note, the envelope of the one-parameter set of quadrics in principal position may have more than one connected component, since not all quadrics have to be real.

2.12 Biarc Construction

Originally, biarcs are circular arcs that interpolate given points P_i with additional tangent information T_i, $i = 0, 1$. At each point, the circular arcs have G^1-continuity, see Fig. 2.9. All transition points are located on a fixed *transition circle* j passing through the points P_0 and P_1. In Fig. 2.9 (left) two possible biarcs c_0 and c_1 for given data (P_i, T_i), $i = 0, 1$ are presented. On the right side of Fig. 2.9 the orientation of the tangent at P_1 is changed an the resulting biarc c_2 has a cusp. A generalization are biarcs on quadrics, *i.e.*, conic sections on a quadric that have G^1-continuity at the transition points. Biarcs on quadrics were investigated by W. WANG and B. JOE, see [64]. With the help of biarcs on quadrics circular arcs in three-dimensional space can be achieved, because two points with skew tangents define a sphere. On this sphere the biarc can be obtained as conic section. The basics of the biarc construction on quadrics can also be found in [56]. Since we want to formulate biarcs in the context of chain geometry we repeat the necessary basics.

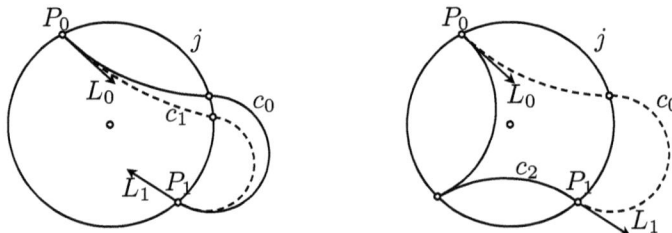

Figure 2.9: Biarcs corresponding to (P_i, L_i), $i = 0, 1$, where L_i is the tangent defined by $[P_i, T_i]$, $i = 0, 1$

The Biarc Construction in projective Space To describe the biarc construction we follow [56]. Let P_0, T_0 and P_1, T_1 be points in projective space written in homogeneous coordinates with respect to some coordinate system:
$$P_0 = p_0 \mathbb{R}, \quad P_1 = p_1 \mathbb{R}, \quad T_0 = t_0 \mathbb{R}, \quad T_1 = t_1 \mathbb{R}.$$
Let κ be the polarity and M the symmetric matrix describing κ. The quadric $\mathfrak{Q}$ is the set of points $X = x\mathbb{R} : x^T M x = 0$. We assume that the points $P_0, P_1 \in \mathfrak{Q}$ and $[P_0, T_0]$, $[P_1, T_1]$ are the tangents of $\mathfrak{Q}$ at the points P_0 and P_1. Thus, we have
$$p_0^T M p_0 = 0, \quad p_1^T M p_1 = 0, \quad p_0^T M t_0 = 0, \quad p_1^T M t_0 = 0.$$
The connection point $C = c\mathbb{R}$ of both conic sections has to lie on $\mathfrak{Q}$: $c^T M c = 0$. The tangents at P_0 and P_1 are given by
$$b = p_0 + \lambda t_0, \quad d = p_1 + \mu t_1, \quad \lambda, \mu \in \mathbb{R}.$$

2.12 Biarc Construction

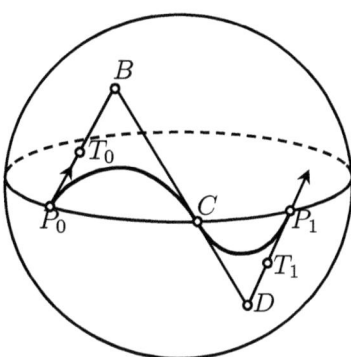

Figure 2.10: Biarc construction on a sphere. The transition point C is located on the conic sections through P_0 and P_1.

To obtain G^1-continuity the point C has to be located on the line spanned by b and d, i.e., $c = \alpha b + \beta d$. Therefore, we can formulate one condition for G^1-continuity. The line joining $B = b\mathbb{R}$ and $D = d\mathbb{R}$ has to be tangent to $\mathfrak{Q}$, see Fig. 2.10. This means it has exactly one intersection point with $\mathfrak{Q}$. Thus, we compute

$$(\alpha b + \beta d)^{\mathrm{T}} \mathrm{M}(\alpha b + \beta d) = \alpha^2 b^{\mathrm{T}} \mathrm{M} b + \beta^2 d^{\mathrm{T}} \mathrm{M} d + \alpha\beta b^{\mathrm{T}} \mathrm{M} d + \beta\alpha d^{\mathrm{T}} \mathrm{M} b \quad (2.18)$$

$$= \alpha^2 b^{\mathrm{T}} \mathrm{M} b + \beta^2 d^{\mathrm{T}} \mathrm{M} d + 2\alpha\beta b^{\mathrm{T}} \mathrm{M} d = 0.$$

As solution of this equation we need a double root for G^1-continuity. This leads to

$$\frac{\alpha^2}{\beta^2} + 2\frac{\alpha}{\beta}\frac{b^{\mathrm{T}} \mathrm{M} d}{b^{\mathrm{T}} \mathrm{M} b} + \frac{d^{\mathrm{T}} \mathrm{M} d}{b^{\mathrm{T}} \mathrm{M} b} = 0 \implies \left(\frac{\alpha}{\beta}\right)_{1,2} = -\frac{\alpha}{\beta}\frac{b^{\mathrm{T}} \mathrm{M} d}{b^{\mathrm{T}} \mathrm{M} b} \pm \sqrt{\frac{(b^{\mathrm{T}} \mathrm{M} d)^2}{(b^{\mathrm{T}} \mathrm{M} b)^2} + \frac{d^{\mathrm{T}} \mathrm{M} d}{b^{\mathrm{T}} \mathrm{M} b}}.$$

To obtain a double root the discriminant has to vanish:

$$\begin{aligned}
0 &= (b^{\mathrm{T}} \mathrm{M} d)^2 + (b^{\mathrm{T}} \mathrm{M} b)(d^{\mathrm{T}} \mathrm{M} d) \qquad (2.19) \\
&= \left((p_0 + \lambda t_0)^{\mathrm{T}} \mathrm{M}(p_0 + \lambda t_0)\right)\left((p_1 + \mu t_1)^{\mathrm{T}} \mathrm{M}(p_1 + \mu t_1)\right) \\
&\quad + \left((p_0 + \lambda t_0)^{\mathrm{T}} \mathrm{M}(p_1 + \mu t_1)\right)^2 \\
&= p_0^{\mathrm{T}} \mathrm{M} p_1 + \mu p_0^{\mathrm{T}} \mathrm{M} t_1 + \lambda t_0^{\mathrm{T}} \mathrm{M} p_1 + \lambda \mu \left(t_0^{\mathrm{T}} \mathrm{M} t_1 \pm \sqrt{t_0^{\mathrm{T}} \mathrm{M} t_0 \cdot t_1^{\mathrm{T}} \mathrm{M} t_1}\right).
\end{aligned}$$

If $t_0^T M t_0 \cdot t_1^T M t_1 \geq 0$ Eq. (2.19) is solvable and we can determine $\lambda : \mu$ and $\alpha : \beta$ afterwards. The set of points that serves as connection point is determined by the two conics through P_0 and P_1 that are solutions of Eq. (2.18). Note, that the existence of a biarc cannot be guaranteed in general. If both tangents are situated on different sides of the quadric, there is no real point $C = c\mathbb{R}$. This is the geometric interpretation of the vanishing discriminant, see Eq. (2.19). If the quadric $\mathfrak{Q}$ is an oval one, there is no tangent in the interior of $\mathfrak{Q}$. Thus, for oval quadrics we can always find biarcs. Naturally, a biarc is not determined uniquely. The set of all possible transition points is contained in two different conics passing through P_0 and P_1, see [56].

2.12.1 Biarcs as touching Chains

Conic sections on arbitrary quadrics can be described with the help of chains in a geometric algebra corresponding to this quadric, see section 2.8. Points are null vectors and the chains can be calculated with Eq. (2.8). Furthermore, we introduced the concept of contact spaces in section 2.6. Touching chains can be interpreted as conics with G^1-continuity. We can use Th. 2.16 to parametrise the chain in contact to another chain at a certain point. To apply this to biarc construction, we have to change the setting. We do not assume that we are given two points and their tangents, since we do not have tangents in the Clifford algebra model. We start with a chain $c_{\mathfrak{abc}}$ determined by the three null vectors $\mathfrak{a}, \mathfrak{b}, \mathfrak{c} \in \bigwedge^1 V$ and compute the chain through a null vector $\mathfrak{d}$ that has contact with $c_{\mathfrak{abc}}$ at $\mathfrak{c}$. The advantage of this method is that we obtain a parametrisation for each biarc. Furthermore, we can use the fact that connected components of the Pin group and the grade-1 subspace form subchain geometries to describe biarc interpolation in an affine Clifford algebra setting.

Biarc Construction with Clifford Algebra Let $c_{\mathfrak{abc}}$ be the chain defined by $\mathfrak{a}, \mathfrak{b}, \mathfrak{c} \in \mathcal{Cl}_{(p,q,r)}$. Let $\mathfrak{d} \in \mathcal{Cl}_{(p,q,r)}$ be an element with $\mathfrak{d} \notin c_{\mathfrak{abc}}$. With Th. 2.16 we can parametrise the chain $c_{\mathfrak{d}}$ that has contact with $c_{\mathfrak{abc}}$ at $\mathfrak{c}$ by

$$cr(\mathfrak{a}, \mathfrak{b}, \mathfrak{s}, \mathfrak{c}) - cr(\mathfrak{a}, \mathfrak{b}, \mathfrak{d}, \mathfrak{c}) = \lambda, \quad \lambda \in \mathcal{C}(\mathcal{Cl}_{p,q,r}),$$

2.12 Biarc Construction

where $\mathfrak{s}$ is contained by $c_{\mathfrak{d}}$. Now we compute $\mathfrak{s}(\lambda)$:

$$cr(\mathfrak{a},\mathfrak{b},\mathfrak{s},\mathfrak{c}) - cr(\mathfrak{a},\mathfrak{c},\mathfrak{d},\mathfrak{b}) = \lambda \qquad (2.20)$$

$$(\mathfrak{c}-\mathfrak{a})(\mathfrak{b}-\mathfrak{a})^{-1}(\mathfrak{b}-\mathfrak{s})(\mathfrak{c}-\mathfrak{s})^{-1} - cr(\mathfrak{a},\mathfrak{b},\mathfrak{d},\mathfrak{c}) = \lambda,$$

$$(\mathfrak{c}-\mathfrak{a})(\mathfrak{b}-\mathfrak{a})^{-1}\mathfrak{b} - (\mathfrak{c}-\mathfrak{a})(\mathfrak{b}-\mathfrak{a})^{-1}\mathfrak{s} = (\lambda + cr(\mathfrak{a},\mathfrak{c},\mathfrak{d},\mathfrak{b}))(\mathfrak{c}-\mathfrak{s}),$$

$$\left[\lambda + cr(\mathfrak{a},\mathfrak{b},\mathfrak{d},\mathfrak{c}) - (\mathfrak{c}-\mathfrak{a})(\mathfrak{b}-\mathfrak{a})^{-1}\right]\mathfrak{s} = \left[\lambda\mathfrak{c} + cr(\mathfrak{a},\mathfrak{b},\mathfrak{d},\mathfrak{c})\mathfrak{c} - (\mathfrak{c}-\mathfrak{a})(\mathfrak{b}-\mathfrak{a})^{-1}\mathfrak{b}\right],$$

$$\mathfrak{s} = \left[\lambda + cr(\mathfrak{a},\mathfrak{b},\mathfrak{d},\mathfrak{c}) - (\mathfrak{c}-\mathfrak{a})(\mathfrak{b}-\mathfrak{a})^{-1}\right]^{-1}\left[\lambda\mathfrak{c} + cr(\mathfrak{a},\mathfrak{b},\mathfrak{d},\mathfrak{c})\mathfrak{c} - (\mathfrak{c}-\mathfrak{a})(\mathfrak{b}-\mathfrak{a})^{-1}\mathfrak{b}\right],$$

with

$$cr(\mathfrak{a},\mathfrak{b},\mathfrak{d},\mathfrak{c}) = (\mathfrak{c}-\mathfrak{a})(\mathfrak{b}-\mathfrak{a})^{-1}(\mathfrak{b}-\mathfrak{d})(\mathfrak{c}-\mathfrak{d})^{-1}.$$

For our purpose the elements $\mathfrak{a},\mathfrak{b},\mathfrak{c},\mathfrak{d}$ are null vectors in a Clifford algebra constructed with the matrix of a polarity of some quadric. Furthermore, we assume $\lambda \in \mathbb{R}$ to obtain conic sections. Formula (2.20) can be implemented in a computer algebra system to obtain parametrisations of the chain that is in contact with $c_{\mathfrak{a}\mathfrak{b}\mathfrak{c}}$ at the element $\mathfrak{c}$. The same calculation can be performed in an affine setting. Therefore, we model the quadric in the grade-1 space with an affine Clifford algebra model. Points of the quadric are represented by elements that square to 1.

2.12.2 Biarcs on Quadrics in three-dimensional Space

Now we give some examples for biarcs on quadrics in three-dimensional space calculated with the help of chains that are in contact.

Sphere The first example is the unit sphere $S^2 \subseteq \mathbb{R}^3$ given by its implicit equation

$$x^2 + y^2 + z^2 = 1.$$

Projective Setting The unit sphere can be considered as projective quadric

$$\mathfrak{Q}: x^T M x = 0, \text{ with } X = x\mathbb{R} \in \mathbb{P}^3(\mathbb{R}) \text{ and } M = \begin{pmatrix} -1 & 0 & 0 & 0 \\ 0 & 1 & 0 & 0 \\ 0 & 0 & 1 & 0 \\ 0 & 0 & 0 & 1 \end{pmatrix}.$$

Thus, we use the homogeneous Clifford algebra model constructed with $V = \mathbb{R}^4$ as vector space model for $\mathbb{P}^3(\mathbb{R})$ together with the quadratic form defined by M. Let $\mathfrak{a} = e_1 + e_4$, $\mathfrak{b} = e_1 + e_2$, and $\mathfrak{c} = e_1 + e_3$ be three mutually distant null vectors. Note that the homogeneous factor is contained in the e_1 generator and that e_2, e_3, e_4 correspond to the x-, y-, z-coordinate. The chain $c_{\mathfrak{abc}}(\lambda)$ is computed with Eq. (2.8):

$$c_{\mathfrak{abc}}(\lambda) = e_1 + \frac{\lambda e_2}{1 - \lambda + \lambda^2} + \frac{\lambda(\lambda-1)e_3}{(1-\lambda+\lambda^2)} + \frac{(1-\lambda)e_4}{1-\lambda+\lambda^2}.$$

Let $\mathfrak{d} = e_1 + \frac{1}{\sqrt{2}}e_2 + \frac{1}{\sqrt{2}}e_3$ be the null vector corresponding to $D = d\mathbb{R} = (1, \frac{1}{\sqrt{2}}, \frac{1}{\sqrt{2}}, 0)^T \mathbb{R}$. We compute the chain that has contact with $c_{\mathfrak{abc}}$ at $\mathfrak{c}$ with Eq. (2.20)

$$\mathfrak{s}(\lambda) = e_1 - \frac{(1+\sqrt{2}-\lambda)e_2}{(\sqrt{2}+1)\lambda - \lambda^2 - \sqrt{2} - 2} + \frac{((1+\sqrt{2})\lambda - \sqrt{2} - \lambda^2 - 1)e_3}{(\sqrt{2}+1)\lambda - \lambda^2 - \sqrt{2} - 2} - \frac{\lambda e_4}{(\sqrt{2}+1)\lambda - \lambda^2 - \sqrt{2} - 2}.$$

This conic section interpolates the point $C = c\mathbb{R}$ related to the null vector $\mathfrak{c}$ at $\lambda = 0$ and the point $D = d\mathbb{R}$ at the parameter $\lambda = \infty$. A visualisation of the biarc construction on the sphere is given in Fig. 2.11 (left).

Affine Setting Now we compute the same biarc with an affine Clifford algebra. Therefore, let $V = \mathbb{R}^3$ and let $Q = \begin{pmatrix} 1 & 0 & 0 \\ 0 & 1 & 0 \\ 0 & 0 & 1 \end{pmatrix}$. The Clifford algebra has signature $(p,q,r) = (3,0,0)$. The square of a general grade-1 element $\mathfrak{a} = a_1 e_1 + a_2 e_2 + a_3 e_3$ is computed as $\mathfrak{a}^2 = a_1^2 + a_2^2 + a_3^2$. If the square is equal to 1 the element corresponds to a point on S^2. The points $A = (0,0,1)^T$, $B = (1,0,0)^T$, and $C = (0,1,0)^T$ that were also used in the example above correspond to the elements $\mathfrak{a} = e_3$, $\mathfrak{b} = e_1$ and $\mathfrak{c} = e_2$ with $\mathfrak{a}^2 = \mathfrak{b}^2 = \mathfrak{c}^2 = 1$. The chain $c_{\mathfrak{abc}}(\lambda)$ defined by these three vectors is given by

$$c_{\mathfrak{abc}}(\lambda) = \frac{\lambda e_1}{1-\lambda+\lambda^2} + \frac{(\lambda-1)\lambda e_2}{1-\lambda+\lambda^2} + \frac{(1-\lambda)e_3}{1-\lambda+\lambda^2}.$$

2.12 Biarc Construction

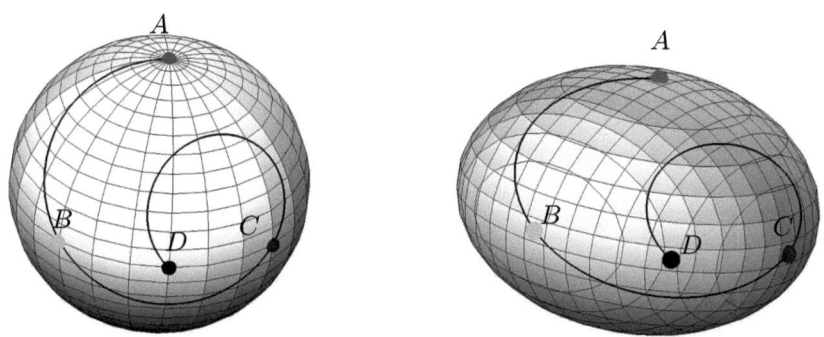

Figure 2.11: Biarcs on a sphere (left); on an ellipsoid (right).

Let $D = (\frac{1}{\sqrt{2}}, \frac{1}{\sqrt{2}}, 0)^T$ be a point on S^2 with algebra representation $\mathfrak{d} = \frac{1}{\sqrt{2}} e_1 + \frac{1}{\sqrt{2}} e_2$. The chain $\mathfrak{s}(\lambda)$ through $\mathfrak{d}$ that has contact with $c_{\mathfrak{abc}}$ at $\mathfrak{c}$ is computed with Eq. (2.20)

$$\mathfrak{s}(\lambda) = -\frac{(1+\sqrt{2}-\lambda)e_1}{(\sqrt{2}+1)\lambda - \lambda^2 - \sqrt{2} - 2} + \frac{((1+\sqrt{2})\lambda - \sqrt{2} - \lambda^2 - 1)e_2}{(\sqrt{2}+1)\lambda - \lambda^2 - \sqrt{2} - 2}$$
$$- \frac{\lambda e_3}{(\sqrt{2}+1)\lambda - \lambda^2 - \sqrt{2} - 2}.$$

If we compare the results of the affine biarc construction with the results of the projective setting, we observe that they are absolutely identical. Therefore, it does not matter which method we take. The affine version is faster because the corresponding algebra has half the dimension of the algebra corresponding to the projective version. In general it is good to have both versions since there are quadrics like Klein's quadric that are naturally projective quadrics.

Ellipsoid The second example we discuss is an ellipsoid with the equation
$$x^2 + y^2 + 2z^2 = 1.$$

Projective Setting The ellipsoid is given by

$$\mathfrak{Q}: x^T M x = 0, \text{ with } X = x\mathbb{R} \in \mathbb{P}^3(\mathbb{R}) \text{ and } M = \begin{pmatrix} -1 & 0 & 0 & 0 \\ 0 & 1 & 0 & 0 \\ 0 & 0 & 1 & 0 \\ 0 & 0 & 0 & 2 \end{pmatrix}.$$

We construct the homogeneous Clifford algebra model with $V = \mathbb{R}^4$ and $Q = M$. Again, the homogeneous factor is stored in the e_1 component and the x-, y-, z-coordinate is contained in e_2, e_3, e_4. Let $A = a\mathbb{R} = (1, 0, 0, \frac{1}{\sqrt{2}})^T \mathbb{R}$, $B = b\mathbb{R} = (1, 1, 0, 0)^T \mathbb{R}$, and $C = c\mathbb{R} = (1, 0, 1, 0)^T \mathbb{R}$ be three points on $\mathfrak{Q}$. The corresponding null vectors are given by $\mathfrak{a} = e_1 + \frac{1}{\sqrt{2}} e_4$, $\mathfrak{b} = e_1 + e_2$, and $\mathfrak{c} = e_1 + e_3$. The chain $c_{\mathfrak{abc}}(\lambda)$ computed via Eq. (2.8) results in

$$c_{\mathfrak{abc}}(\lambda) = e_1 + \frac{\lambda e_2}{1 - \lambda + \lambda^2} + \frac{\lambda(\lambda - 1)e_3}{1 - \lambda + \lambda^2} + \frac{1}{2} \frac{\sqrt{2}(1 - \lambda)e_4}{1 - \lambda + \lambda^2}.$$

Let $D = d\mathbb{R} = (1, \frac{1}{\sqrt{2}}, \frac{1}{\sqrt{2}}, 0)^T \mathbb{R}$ be an additional point on $\mathfrak{Q}$ with algebra representation $\mathfrak{d} = e_1 + \frac{1}{\sqrt{2}} e_2 + \frac{1}{\sqrt{2}} e_3$. The chain $c_{\mathfrak{d}}(\lambda)$ that has contact with $c_{\mathfrak{abc}}(\lambda)$ is calculated with the help of Eq. (2.20) as:

$$c_{\mathfrak{d}}(\lambda) = e_1 + \frac{(\lambda - \sqrt{2} - 1)e_2}{\lambda + \sqrt{2}\lambda - \lambda^2 - 2 - \sqrt{2}} + \frac{(\sqrt{2}\lambda - 1 - \sqrt{2} - \lambda^2 + \lambda)e_3}{\lambda + \sqrt{2}\lambda - \lambda^2 - 2 - \sqrt{2}}$$
$$- \frac{1}{2} \frac{\sqrt{2}\lambda e_4}{\lambda + \sqrt{2}\lambda - \lambda^2 - 2 - \sqrt{2}}.$$

Affine Setting For the affine biarc construction on the ellipsoid we use the Clifford algebra constructed by $V = \mathbb{R}^3$ together with the quadratic form $Q = \begin{pmatrix} 1 & 0 & 0 \\ 0 & 1 & 0 \\ 0 & 0 & 2 \end{pmatrix}$. The points $A = (0, 0, \frac{1}{\sqrt{2}})^T$, $B = (1, 0, 0)^T$, and $C = (0, 1, 0)^T$ on the ellipsoid correspond to elements $\mathfrak{a} = \frac{1}{\sqrt{2}} e_2$, $\mathfrak{b} = e_1$, and $\mathfrak{c} = e_2$, with $\mathfrak{a}^2 = \mathfrak{b}^2 = \mathfrak{c}^2 = 1$. The chain $c_{\mathfrak{abc}}(\lambda)$ is determined as

$$c_{\mathfrak{abc}}(\lambda) = \frac{\lambda e_1}{1 - \lambda + \lambda^2} + \frac{\lambda(\lambda - 1)e_2}{1 - \lambda + \lambda^2} + \frac{1}{2} \frac{\sqrt{2}(1 - \lambda)e_3}{1 - \lambda + \lambda^2}.$$

2.12 Biarc Construction

Let $D = (\frac{1}{\sqrt{2}}, \frac{1}{\sqrt{2}}, 0)^T$ be another point on $\mathfrak{Q}$ with algebra representation $\mathfrak{d} = \frac{1}{\sqrt{2}}e_1 + \frac{1}{\sqrt{2}}e_2$. The chain in contact with $c_{\mathfrak{abc}}$ that contains $\mathfrak{d}$ is computed via Eq. (2.20)

$$c_{\mathfrak{d}}(\lambda) = \frac{(\lambda - \sqrt{2} - 1)e_1}{\lambda + \sqrt{2}\lambda - \lambda^2 - 2 - \sqrt{2}} + \frac{(\sqrt{2}\lambda - 1 - \sqrt{2} - \lambda^2 + \lambda)e_2}{\lambda + \sqrt{2}\lambda - \lambda^2 - 2 - \sqrt{2}}$$
$$- \frac{1}{2}\frac{\sqrt{2}\lambda e_3}{\lambda + \sqrt{2}\lambda - \lambda^2 - 2 - \sqrt{2}}.$$

The biarc construction performed above is displayed in Fig 2.11 (right).

One sheeted Hyperboloid Now we do the same on the one sheeted hyperboloid given by the equation

$$x^2 + y^2 - z^2 = 1.$$

For quadrics that contain ideal points, it is advantageous to perform the construction in the projective setting, because in this setting it possible to describe also ideal elements. Therefore, we do not use the affine setting for this example.

Projective Setting The Clifford algebra model is constructed with $V = \mathbb{R}^4$ as vector space model for $\mathbb{P}^3(\mathbb{R})$ and the quadratic form $Q = M$ corresponding to the quadric

$$\mathfrak{Q} : x^T M x = 0, \text{ with } X = x\mathbb{R} \in \mathbb{P}^3(\mathbb{R}) \text{ and } M = \begin{pmatrix} -1 & 0 & 0 & 0 \\ 0 & 1 & 0 & 0 \\ 0 & 0 & 1 & 0 \\ 0 & 0 & 0 & -1 \end{pmatrix}.$$

We start with the points $A = a\mathbb{R} = (1, \frac{1}{\sqrt{2}}, \frac{1}{\sqrt{2}}, 0)^T\mathbb{R}$, $B = b\mathbb{R}(1, 1, 0, 0)^T\mathbb{R}$, and $C = c\mathbb{R}(1, 0, -1, 0)^T\mathbb{R} \in \mathfrak{Q}$. The corresponding algebra elements are given by $\mathfrak{a} = e_1 + \frac{1}{\sqrt{2}}e_2 + \frac{1}{\sqrt{2}}e_3$, $\mathfrak{b} = e_1 + e_2$, $\mathfrak{c} = e_1 - e_3$. We compute the chain defined by these three null vectors as

$$c_{\mathfrak{abc}}(\lambda) = e_1 + \frac{(2+\sqrt{2})\lambda + 1 + \sqrt{2}}{\sqrt{2} + \lambda\sqrt{2} + \lambda^2 + 2}e_2 - \frac{\lambda^2 + \lambda\sqrt{2} - 1 - \sqrt{2}}{\sqrt{2} + \lambda\sqrt{2} + \lambda^2 + 2}e_3.$$

The next point that shall be interpolated is $D = d\mathbb{R} = (1, -2, 1, 2)^T\mathbb{R} \in$

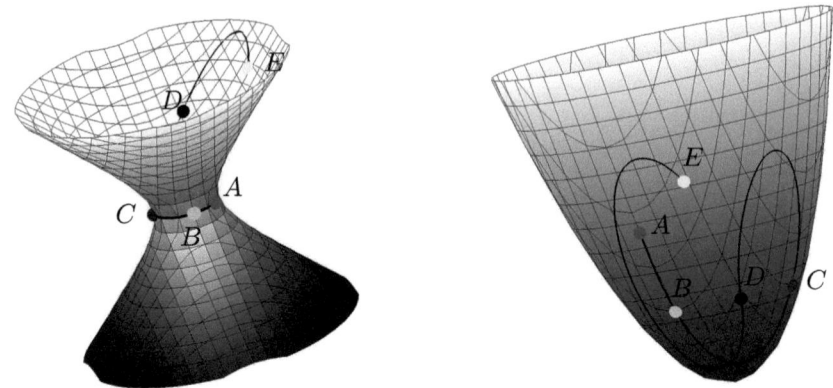

Figure 2.12: Biarc on an one sheeted hyperboloid (left); on a paraboloid (right).

$\mathfrak{Q}$. Its algebra representation is $\mathfrak{d} = e_1 - 2e_2 + e_3 + 2e_4$ and the chain through $\mathfrak{d}$ in contact with $c_{\mathfrak{abc}}$ at $\mathfrak{c}$ results in

$$c_{\mathfrak{d}}(\lambda) = \frac{2(3 + 4\lambda + 2\lambda^2 + 2\lambda\sqrt{2} + 2\sqrt{2})e_1}{(2 + \sqrt{2} + 2\lambda)^2} - \frac{4(2\sqrt{2} + 3 + \lambda\sqrt{2} + 2\lambda)e_2}{(2 + \sqrt{2} + 2\lambda)^2}$$
$$- \frac{2(-2\sqrt{2} - 3 + 4\lambda + 2\lambda^2 + 2\lambda\sqrt{2})e_3}{(2 + \sqrt{2} + 2\lambda)^2} + \frac{4(2\sqrt{2} + 3)e_4}{(2 + \sqrt{2} + 2\lambda)^2}.$$

Note that $c_{\mathfrak{d}}$ contains a null vector corresponding to an ideal point for the parameter value $\lambda = -1 - \frac{1}{2}\sqrt{2}$. Let $E = e\mathbb{R} = (1, -1, 3, 3)^T\mathbb{R}$ be another point on $\mathfrak{Q}$ with algebra representation $\mathfrak{e} = e_1 - e_2 + 3e_3 + 3e_4$. The chain $c_{\mathfrak{e}}$ in contact with $c_{\mathfrak{d}}$ at $\mathfrak{d}$ has the form:

$$c_{\mathfrak{e}}(\lambda) = e_1 - \frac{2(-12 + 8\sqrt{2} - \lambda^2 + (4\sqrt{2} - 8)\lambda)e_2}{-\lambda^2 + (2\sqrt{2} - 4)\lambda - 24 + 16\sqrt{2}}$$
$$+ \frac{(-72 + 48\sqrt{2} - \lambda^2 + (10\sqrt{2} - 20)\lambda)e_3}{-\lambda^2 + (2\sqrt{2} - 4)\lambda - 24 + 16\sqrt{2}}$$
$$+ \frac{2(-36 + 24\sqrt{2} - \lambda^2 + (6\sqrt{2} - 12)\lambda)e_4}{-\lambda^2 + (2\sqrt{2} - 4)\lambda - 24 + 16\sqrt{2}}.$$

The corresponding biarc interpolation is presented in Fig. 2.12 (left).

2.12 Biarc Construction

Paraboloid As last example we examine the biarc construction on a paraboloid with equation:

$$x^2 + y^2 - z = 0.$$

Projective Setting The homogeneous Clifford algebra model is constructed with $V = \mathbb{R}^4$ as model for $\mathbb{P}^3(\mathbb{R})$ together with the quadratic form Q with the coefficient matrix

$$\mathfrak{Q}: x^T M x = 0, \text{ with } X = x\mathbb{R} \in \mathbb{P}^3(\mathbb{R}) \text{ and } M = \begin{pmatrix} 0 & 0 & 0 & \frac{1}{2} \\ 0 & 1 & 0 & 0 \\ 0 & 0 & 1 & 0 \\ \frac{1}{2} & 0 & 0 & 0 \end{pmatrix}.$$

As example we again choose three points on the paraboloid. Let $A = a\mathbb{R} = (1, \sqrt{2}, 0, 2)^T \mathbb{R}$, $B = b\mathbb{R} = (1, 1, 0, 1)^T \mathbb{R}$, and $C = c\mathbb{R} = (1, 0, 1, 1)^T \mathbb{R}$ be three different points with algebra representation $\mathfrak{a} = e_1 + \sqrt{2}e_2 + 2e_4$, $\mathfrak{b} = e_1 + e_2 + e_4$, $\mathfrak{c} = e_1 + e_3 + e_4$. We compute the chain $c_{\mathfrak{abc}}$ with Eq. (2.8).

$$c_{\mathfrak{abc}}(\lambda) = e_1 + \frac{(\lambda + 8 + 6\sqrt{2})e_2}{\lambda^2 + (2\sqrt{2}+2)\lambda + 6 + 4\sqrt{2}} + \frac{\lambda(-1+\lambda)e_3}{\lambda^2 + (2\sqrt{2}+2)\lambda + 6 + 4\sqrt{2}} - \frac{(-\lambda^2 + (2\sqrt{2}+4)\lambda - 12 - 8\sqrt{2})e_4}{\lambda^2 + (2\sqrt{2}+2)\lambda + 6 + 4\sqrt{2}}.$$

Let $D = d\mathbb{R} = (1, \frac{1}{\sqrt{2}}, \frac{1}{\sqrt{2}}, 1)^T \mathbb{R}$ and $E = e\mathbb{R} = (1, \frac{2}{\sqrt{2}}, \frac{1}{\sqrt{2}}, \frac{5}{2})^T \mathbb{R}$ be two additional points on the paraboloid and $\mathfrak{d} = e_1 + \frac{1}{\sqrt{2}}e_2 + \frac{1}{\sqrt{2}}e_3 + e_4$, $\mathfrak{e} = e_1 + \frac{2}{\sqrt{2}}e_2 + \frac{1}{\sqrt{2}}e_3 + \frac{5}{2}e_4$ the corresponding null vectors. The chain $c_\mathfrak{d}$ that is in contact with $c_{\mathfrak{abc}}$ at $\mathfrak{c}$ is computed with Eq. (2.20) and has the form

$$c_\mathfrak{d}(\lambda) = e_1 - \frac{(-\lambda + 21 + 15\sqrt{2})e_2}{-\lambda^2 + (3\sqrt{2}+4)\lambda - 21\sqrt{2} - 30} + \frac{(-\lambda^2 + (\sqrt{2}+1)\lambda - 21 - 15\sqrt{2})e_3}{-\lambda^2 + (3\sqrt{2}+4)\lambda - 21\sqrt{2} - 30} - \frac{(\lambda^2 + (\sqrt{2}+2)\lambda + 30 + 21\sqrt{2})e_4}{-\lambda^2 + (3\sqrt{2}+4)\lambda - 21\sqrt{2} - 30}.$$

Moreover, we compute the chain $c_\mathfrak{e}$ that has contact with $c_\mathfrak{d}$ at $\mathfrak{d}$

$$c_\mathfrak{e}(\lambda) = e_1 + \frac{1}{2}\frac{(-384 - 36\lambda\sqrt{2} + 3\lambda^2\sqrt{2} + 272\sqrt{2} + 48\lambda)e_2}{3\lambda^2 + (16\sqrt{2} - 24)\lambda - 96\sqrt{2} + 136}$$
$$+ \frac{1}{2}\frac{((20 - 16\sqrt{2})\lambda + 3\sqrt{2}\lambda^2 + 136\sqrt{2} - 192)e_3}{3\lambda^2 + (16\sqrt{2} - 24)\lambda - 96\sqrt{2} + 136}$$
$$+ \frac{(3\lambda^2 + (18\sqrt{2} - 28)\lambda + 340 - 240\sqrt{2})e_4}{3\lambda^2 + (16\sqrt{2} - 24)\lambda - 96\sqrt{2} + 136}.$$

An illustration of the paraboloid together with the conic section belonging to $c_{\mathfrak{abc}}, c_\mathfrak{d}$ and $c_\mathfrak{e}$ is given in Fig. 2.12 (right).

2.12.3 Klein's Quadric

To present an example in a higher-dimensional projective space we construct biarcs on Klein's quadric M_2^4. We introduced Klein's quadric in section 1.2.1. Conics on M_2^4 correspond to reguli. Thus, a biarc construction on M_2^4 can directly be transferred to an interpolation algorithm that interpolates a set of given lines by reguli. To demonstrate the power of the Clifford algebra apparatus we apply the concept of touching chains to the homogeneous Clifford algebra model for Klein's quadric, see section 1.6. Hence, we use $V = \mathbb{R}^6$ as model for $\mathbb{P}^5(\mathbb{R})$ and

$$Q = \begin{pmatrix} O & I \\ I & O \end{pmatrix},$$

where O denotes 3×3 zero matrix and I the 3×3 identity matrix. As example we take the same conic as in Ex. 2.5, but with another parametrisation. We start with the Plücker coordinates $A = a\mathbb{R} = (0,1,0,0,0,1)^T\mathbb{R}$, $B = b\mathbb{R} = (-1,1,1,0,0,0)^T\mathbb{R}$, and $C = c\mathbb{R} = (0,0,1,0,1,0)^T\mathbb{R}$. Obviously, $A, B, C \in M_2^4$. The corresponding null vectors are given by

$$\mathfrak{a} = e_2 + e_6, \quad \mathfrak{b} = -e_1 + e_2 + e_3, \quad \mathfrak{c} = e_3 + e_5.$$

The chain defined by these three null vectors is computed with Eq. (2.8)

$$c_{\mathfrak{abc}}(\lambda) = \frac{-2\lambda e_1}{1+\lambda^2} + \frac{(1+\lambda)e_2}{1+\lambda^2} + \frac{\lambda(1+\lambda)e_3}{1+\lambda^2} + \frac{\lambda(\lambda-1)e_5}{1+\lambda^2} - \frac{(\lambda-1)e_6}{1+\lambda^2}.$$

2.12 Biarc Construction

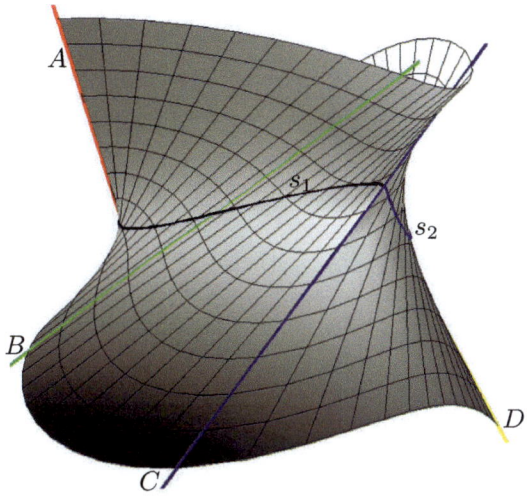

Figure 2.13: Biarc construction with reguli.

Let $D = d\mathbb{R} = (552, -140, 371, -336, 827, 812)^T \mathbb{R}$ be another point on M_2^4 with algebra representation

$$\mathfrak{d} = 552e_1 - 140e_2 + 371e_3 - 336e_4 + 827e_5 + 812e_6.$$

The chain $c_\mathfrak{d}$ containing $\mathfrak{d}$ that has contact with $c_{\mathfrak{abc}}$ is given by:

$$c_\mathfrak{d}(\lambda) = \frac{24(23+28\lambda)e_1}{607+1040\lambda+336\lambda^2} - \frac{28(12\lambda+5)e_2}{607+1040\lambda+336\lambda^2} + \frac{(704\lambda+336\lambda^2+371)e_3}{607+1040\lambda+336\lambda^2}$$
$$-\frac{336e_4}{607+1040\lambda+336\lambda^2} + \frac{(1376\lambda+336\lambda^2+827)e_5}{607+1040\lambda+336\lambda^2} + \frac{28(12\lambda+29)e_6}{607+1040\lambda+336\lambda^2}.$$

Fig. 2.13 shows the lines corresponding to the points A, B, C, D, the reguli corresponding to the conic sections, and their striction curves s_1, s_2. These two reguli posses G^1-continuity at the transition line given by its Plücker coordinates C.

2.12.4 Biarcs on the Dual Sphere

With the concept of touching chains we are also able to compute conics on $S_\mathbb{D}^2$ that are in contact, *i.e.*, congruences of lines having a constant dual angle with respect to a fixed line. In general, two such congruences with constant dual angles with respect to a fixed line may have zero, one, two, or infinitely many oriented lines in common. The adequate affine model is the Clifford algebra $\mathcal{C}\ell_{(0,3,0)}(D)$ given by $\mathcal{M} = D^3$ and $Q = \begin{pmatrix} 1 & 0 & 0 \\ 0 & 1 & 0 \\ 0 & 0 & 1 \end{pmatrix}$. As example we give the Plücker coordinates of four lines $A = a\mathbb{R} = (0,1,0,0,0,1)^T\mathbb{R}$, $B = b\mathbb{R} = (-\frac{1}{2}, \frac{2}{3}, \frac{2}{3}, \frac{4}{9}, \frac{1}{9}, \frac{1}{9})^T\mathbb{R}$, $C = c\mathbb{R} = (0,0,1,0,1,0)^T\mathbb{R}$, $D = d\mathbb{R} = (\frac{3}{5}, 0, \frac{4}{5}, -\frac{8}{5}, 2, \frac{6}{5})^T\mathbb{R}$. The corresponding algebra representations where we choose the orientation of the related spears are

$$\mathfrak{a} = e_2 + \epsilon e_3, \qquad \mathfrak{b} = (-\frac{1}{3} + \frac{4}{9}\epsilon)e_1 + (\frac{2}{3} + \frac{1}{9}\epsilon)e_2 + (\frac{2}{3} + \frac{1}{9}\epsilon)e_3,$$

$$\mathfrak{c} = \epsilon e_2 + e_3, \qquad \mathfrak{d} = (\frac{3}{5} - \epsilon\frac{8}{5})e_1 + (\frac{4}{5} + 2\epsilon)e_2 + (\frac{1}{5} + \epsilon\frac{6}{5})e_3.$$

We compute the chains $c_{\mathfrak{abc}}$ and $c_{\mathfrak{acd}}$ for a dual cross ratio with Eq. (2.8).

$$c_{\mathfrak{abc}}(\lambda, \mu) = \left(-\frac{\lambda}{\lambda^2+1+\lambda} + \frac{(\lambda+1)(\lambda^2+\lambda+\mu\lambda-\mu)}{(\lambda^2+1+\lambda)^2}\epsilon \right) e_1$$
$$+ \left(\frac{\lambda+1}{\lambda^2+1+\lambda} + \frac{\lambda(-2\mu-\mu\lambda+\lambda^3+\lambda^2-1)}{(\lambda^2+1+\lambda)^2}\epsilon \right) e_2$$
$$+ \left(\frac{(\lambda+1)\lambda}{\lambda^2+1+\lambda} - \frac{-2\mu\lambda-\mu+\lambda^3-1-\lambda}{(\lambda^2+1+\lambda)^2}\epsilon \right) e_3,$$

$$c_{\mathfrak{acd}}(\lambda, \mu) = \left(\frac{3(\lambda-1)\lambda}{1+5\lambda^2-\lambda} - \frac{-6\mu\lambda+3\mu+4\lambda^2-12\lambda^2\mu-48\lambda^3+40\lambda^4+4\lambda}{(1+5\lambda^2-\lambda)^2}\epsilon \right) e_1$$
$$+ \left(-\frac{\lambda-1}{(1+5\lambda^2-\lambda)} + \frac{\lambda(-10\mu-9\lambda+5\mu\lambda-15\lambda^2+50\lambda^3-1)}{(1+5\lambda^2-\lambda)^2}\epsilon \right) e_2$$
$$+ \left(\frac{(4\lambda+1)\lambda}{(1+5\lambda^2-\lambda)} + \frac{(\lambda-1)(30\lambda^3-31\lambda^2+7\lambda-9\mu\lambda-\mu-1)}{(1+5\lambda^2-\lambda)^2}\epsilon \right) e_3.$$

These two chains parametrise the congruences of lines that corresponds to the intersection of two affine submodules spanned by $\mathfrak{a}, \mathfrak{b}, \mathfrak{c}$ and $\mathfrak{a}, \mathfrak{c}, \mathfrak{d}$ with the dual unit sphere S_D^2. The congruences of lines have exactly two oriented lines in common, i.e., the lines defined by $\mathfrak{a}$ and $\mathfrak{c}$. To construct a biarc we use again the chain geometric method. The

2.12 Biarc Construction

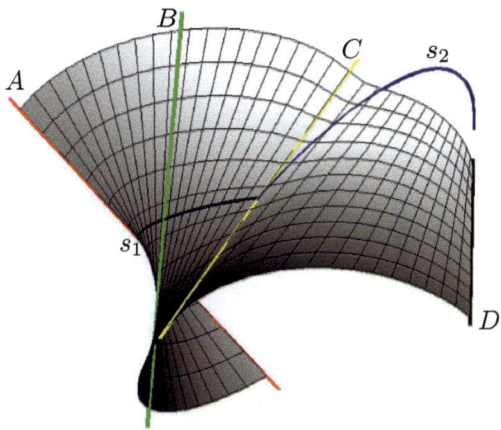

Figure 2.14: Biarc construction on dual unit sphere

chain $c_{\mathfrak{d}}$ that is in contact with $c_{\mathfrak{abc}}$ can be computed with a dual cross ratio $\lambda + \epsilon \mu$, since the scalar domain of the algebra is the commutative ring of dual numbers. Both chains correspond to congruences of lines with constant dual angle to an uniquely determined oriented line. Hence, the two congruences of lines define two cylinders that intersect each other in a single point if the congruences of lines are in contact. Furthermore both congruences of lines have a whole bundle of parallel lines in common. This is obvious because we can translate the line that belongs to both congruences and that passes through the point of contact of the two cylinders along the axis of one of the cylinders and it is still contained in both congruences. A parametrisation of this congruence can be obtained with Eq. (2.20):

$$\begin{aligned}
c_{\mathfrak{d}}(\lambda, \mu) = & \left(\frac{(\lambda + 3)}{5 + 3\lambda + \lambda^2} - \frac{4\mu + 16\lambda + 40 + 6\mu\lambda + \mu\lambda^2 + \lambda^3}{(5 + 3\lambda + \lambda^2)^2} \epsilon \right) e_1 \\
+ & \left(\frac{-\lambda}{5 + 3\lambda + \lambda^2} + \frac{28\lambda^2 - 5\mu + 70\lambda + 50 + \mu\lambda^2 + 7\lambda^3 + \lambda^4}{(5 + 3\lambda + \lambda^2)^2} \epsilon \right) e_2 \\
+ & \left(\frac{\lambda^2 + 3\lambda + 4}{5 + 3\lambda + \lambda^2} + \frac{2\mu\lambda + 5\lambda^2 + 3\mu + 12\lambda + 30 + \lambda^3}{(5 + 3\lambda + \lambda^2)^2} \epsilon \right) e_3.
\end{aligned}$$

This is a parametrisation of the entire congruence. Now we want to examine the same problem with a real value for the cross ratio.

This can be achieved easily by setting $\mu = 0$. The resulting chains correspond to ruled surfaces of constant slope that have G^1-continuity at the transition line. In Fig. 2.14 the ruled surfaces of constant slope corresponding to the biarc are visualized together with the lines corresponding to A, B, C, D and the striction curves of the ruled surfaces s_1, s_2. Thus, we can use biarcs on the dual sphere to perform G^1-interpolation of oriented lines. A parametrisation of the form $f(u,v) = c(u) + v \cdot r(u)$ where $c(u)$ is a curve on the surface and $r(u)$ a curve on the unit sphere $S^2 \subseteq \mathbb{R}^3$ can be computed with the help of the corresponding Plücker coordinates. For the chain $c_{\mathfrak{abc}}(\lambda, 0)$ we obtain the parametrisation

$$f_1(u,v) = \frac{1}{u^2+u+1}\left[\begin{pmatrix} 1-u^2 \\ u \\ -u \end{pmatrix} + v \begin{pmatrix} -u \\ u+1 \\ (u+1)u \end{pmatrix}\right],$$

with striction curve

$$s_1(u) = \frac{1}{u^2+u+1}(1-u^2,\ u,\ -u)^{\mathrm{T}}.$$

For the chain $c_{\mathfrak{d}}(\lambda, 0)$ we obtain

$$f_2(u,v) = \frac{1}{(u^2+3u+5)^2}\left[\begin{pmatrix} -u^4-7u^3-28u^2-68u-40 \\ -u^3-u^2-20u-50 \\ u^3+6u^2+26u+30 \end{pmatrix}\right.$$
$$\left. + v \begin{pmatrix} 15+14u+u^3+6u^2 \\ -5u-u^3-3u^2 \\ 20+27u+6u^3+u^4+18u^2 \end{pmatrix}\right],$$

with striction curve

$$s_2(u) = \frac{1}{u^2+3u+5}(1-u^2-4u,\ -u-10,\ 13u+18)^{\mathrm{T}}.$$

Remark 2.39. *The concept of chains that are in contact is not restricted to the grade-1 space. We used contact chains to construct biarcs on quadrics as they are understood classically. Naturally, chains in the whole algebra can be in contact, and therefore, we can also use this concept to construct chains in subgroups that are in contact with each other. For example we can determine chains in contact within Pin or Spin groups.*

3 Kinematic Mappings for Spin Groups

In this chapter we define Cayley-Klein spaces and show how to describe certain Cayley-Klein spaces within the homogeneous Clifford algebra model. This construction is accomplished in detail for the three-dimensional Euclidean space. Furthermore, the kinematic mapping of Study and the mapping of Blaschke and Grünwald are constructed in a unified method. Matrices of the collineations in the image and pre-image space are derived. The construction is accomplished in detail for the Euclidean spaces of dimension two and three. After that, we give an overview of possible kinematic mappings for Cayley-Klein spaces of dimension two and three. Moreover, the mapping for the four-dimensional Euclidean space is presented. This chapter is already published, see [41].

3.1 Cayley-Klein Geometries and the homogeneous Model

Cayley-Klein spaces are metric spaces constructed within an n-dimensional projective space $\mathbb{P}^n(\mathbb{R})$ with a distinguished quadratic hypersurface given by a quadratic form

$$x^\mathrm{T} \mathrm{M} x = 0,$$

where M is a $(n+1) \times (n+1)$ symmetric matrix and $x \in \mathbb{P}^n$. This idea goes back to A. CAYLEY and F. KLEIN. Comprehensive work on this field of geometry can be found in [44, 52] and [60]. With Silvester's law of inertia we can always find a diagonal matrix corresponding to the quadratic hypersurface, see section 1.3.1. The Cayley-Klein construction provides models for the Euclidean, hyperbolic, elliptic, and many other geometries. An exhaustive treatise of this topic can be found in [25]. The definition of a Cayley-Klein space that we will use can also be found therein.

Definition 3.1. *Let $\mathbb{P}^n(\mathbb{R})$ be the n-dimensional projective space, then $\mathcal{F} \subset \mathbb{P}^n(\mathbb{C})$ is defined by*

$$\mathcal{F}: Q^n_{r_0,q_0} \supset A^{n_1} \supset Q^{n_1-1}_{r_1,q_1} \supset \ldots \supset A^{n_\rho} \supset Q^{n_\rho-1}_{r_\rho,q_\rho}.$$

Here r_i is the rank, q_i the index and $n_i - 1$ the dimension of the quadratic variety $Q^{n_i-1}_{r_i,q_i}$ that is for $i < \rho$ a singular quadric and for $i = \rho$ a regular one. Furthermore, the identity $n + 1 = r_0 + r_1 + r_\rho$ holds. $\mathbb{P}^n$ with absolute figure $\mathcal{F}$ is called a Cayley-Klein space. *The points contained in $\mathcal{F}$ are called* ideal *points and the points in $\mathbb{P}^n \setminus \mathcal{F}$* proper *points. $Q^{n_i-1}_{r_i,q_i}$ is called* absolute quadric *or* absolute cone. *A^{n_i} is called* absolute n_i-subspace. *A Cayley-Klein space where the set of points W is removed is called* sliced *along W.*

Remark 3.1. *Points that are neither ideal nor proper are called* improper *points. These points occur for example in the case of hyperbolic geometry.*

The construction of a Clifford algebra over a projective space results in the so-called *homogeneous model*, see section 1.4.

3.1.1 Cayley-Klein Spaces

At first we present an example for a planar Cayley-Klein space. This means we construct the homogeneous model over $\mathbb{R}^3$ with coordinates x_0, x_1, x_2 as vector space model for $\mathbb{P}^2(\mathbb{R})$ and an absolute figure. We start with the *Euclidean plane*. For this purpose an absolute figure of the form

$$\mathcal{F}: Q^1_{1,0} \supset A^1 \supset Q^0_{2,0} \tag{3.1}$$

is prescribed. The quadric can be written as $Q^1_{1,0} : x_0^2 = 0$. It is singular, thus cone-shaped and has the vertex space $A^1 : x_0 = 0$ which is a line. The regular quadric at the end of the chain is given by $Q^0_{2,0} : x_1^2 + x_2^2 = 0$. The projective automorphisms of $\mathcal{F}$ form a subgroup $\mathrm{PGL}(\mathbb{P}^2, \mathcal{F}) \subset \mathrm{PGL}(\mathbb{P}^2)$ and thus they constitute the group of isometries in this Cayley-Klein geometry. In the Euclidean the subgroup of direct isometries forms the group SE(2) of planar Euclidean displacements. Now we construct the homogeneous model for this Cayley-Klein space. Therefore, we take $\mathbb{R}^3$ as vector space and a quadratic form with signature $(2,0,1)$ to construct the Clifford algebra $\mathcal{C}\ell_{(2,0,1)}$. Points $P = (x_0, x_1, x_2)^T \in \mathbb{P}^2(\mathbb{R})$ of the Euclidean plane are described in the $\bigwedge^2 V$ subspace of the algebra via $\mathfrak{p} = x_0 e_{12} + x_1 e_{13} + x_2 e_{23}$. Note that the homogenizing coordinate of the point is contained in the e_{12} component. The scalar product is calculated by $\mathfrak{p} \cdot \mathfrak{p} = -x_0^2$. Hence, we have the possibility to norm with operations in the algebra. Thus, we find Euclidean geometry considered as a Cayley-Klein geometry within this model. The group $\text{PGL}(\mathbb{P}^2, \mathcal{F})$ can be found in the Clifford algebra as the group $\text{Pin}_{(2,0,1)}$. In fact the Pin group is a double cover of the group $\text{PGL}(\mathbb{P}^2, \mathcal{F})$.

Three-dimensional Euclidean Space The three-dimensional Euclidean space is constructed in the same way. We start with $\mathbb{P}^3(\mathbb{R})$ and the absolute figure

$$\mathcal{F}: Q_{1,0}^2 \supset A^2 \supset Q_{2,0}^1, \qquad (3.2)$$

wherein the first singular quadric in the chain is given by $Q_{1,0}^2 : x_0^2 = 0$ and its vertex is given by $A^2 : x_0 = 0$. Contained in A^2 we have the so called *absolute circle* $Q_{2,0}^1 : x_1^2 + x_2^2 + x_3^2 = 0$. The subgroup $\text{PGL}(\mathbb{P}^3, \mathcal{F}) \subset \text{PGL}(\mathbb{P}^3)$ that fixes this absolute figure can be identified as the group of spatial Euclidean displacements SE(3). We aim at a representation of this Cayley-Klein geometry with a homogeneous Clifford algebra model. Therefore, we take $\mathbb{R}^4$ as vector space model for $\mathbb{P}^3(\mathbb{R})$ together with a quadratic form of signature $(3,0,1)$ to obtain the Clifford algebra $\mathcal{C}\ell_{(3,0,1)}$. The group $\text{Spin}_{(3,0,1)}$ is a double cover of the group SE(3) of Euclidean motions, see for example [59].

Cayley-Klein Geometry via homogeneous Clifford Algebra Models In general, a Cayley-Klein space can be described using the homogeneous Clifford algebra approach. All we need is the vector space model of a projective space and a quadratic form given by its signature. Of course, not every Cayley-Klein space can be described in this way, but for our purposes this is convenient. The three-dimensional Gallilei space considered as Cayley-Klein space serves as counter example

$$\mathcal{F}: Q_{1,0}^2 \supset A^2 \supset Q_{1,0}^1 \supset A^1 \supset Q_{2,0}^0.$$

The quadrics and the subspaces can be described by

$$Q_{1,0}^2 : x_0^2 = 0, \ A^2 : x_0 = 0, \ Q_{1,0}^1 : x_1^2 = 0, \ A^1 : x_1 = 0, \ Q_{2,0}^0 : x_2^2 + x_3^2 = 0.$$

It is not possible to describe this absolute figure with one quadratic form, and therefore, we conclude that there exists no homogeneous Clifford algebra representation for the Gallilei geometry.

3.1.2 A homogeneous Model for Euclidean Geometry

Now we construct the Clifford algebra for the three-dimensional Euclidean space. Therefore, we take $V = \mathbb{R}^4$ as model for the three-dimensional projective space $\mathbb{P}^3(\mathbb{R})$. The signature of the Clifford algebra is $(3,0,1)$. Note that in the literature the signature $(0,3,1)$ is often used, because of the connection to dual quaternions, see [59]. We will do the construction in a more natural way and take the scalar product of Euclidean geometry. We have four basis elements e_1, e_2, e_3, e_4 with $e_1^2 = e_2^2 = e_3^2 = 1$ and $e_4^2 = 0$. The grade-1 space is equipped with the quadratic form, therefore the points are described by grade-3 elements. This approach is called a plane-based approach, see [27]. Grade-2 and grade-1 elements correspond to lines and planes, respectively. Note that not every grade-2 element corresponds to a line. The even part of the algebra $\mathcal{C}\ell_{(3,0,1)}^+$ is spanned by $e_0, e_{12}, e_{13}, e_{14}, e_{23}, e_{24}, e_{34}, e_{1234}$ and the Spin group is a double cover of SE(3). Note that in this case we have $\mathcal{C}\ell_{(3,0,1)}^+ \cong \mathcal{C}\ell_{(0,3,1)}^+$.

Isomorphism to $\mathbb{H}_d$ Dual quaternions were introduced in section 1.1.2. A more detailed introduction can be found in [38]. We give an isomorphism between $\mathcal{C}\ell_{(3,0,1)}^+$ and $\mathbb{H}_d$ as introduced by E. STUDY, cf. [61]. This isomorphism can also be found in [59] and is given by its action on the basis:

$$e_0 \mapsto 1, \quad e_{23} \mapsto \mathbf{i}, \quad e_{31} \mapsto \mathbf{j}, \quad e_{12} \mapsto \mathbf{k}, \quad (3.3)$$
$$-e_{1234} \mapsto \epsilon, \quad e_{14} \mapsto \epsilon\mathbf{i}, \quad e_{24} \mapsto \epsilon\mathbf{j}, \quad e_{34} \mapsto \epsilon\mathbf{k}.$$

We can define a group isomorphism between the group of dual unit quaternions and $\text{Spin}_{(3,0,1)}$ if we restrict the isomorphism (3.3) to one of these groups in the image or pre-image.

3.2 Kinematic Mappings

In this section we give a short introduction to the concept of kinematic mappings. We will treat two important examples. Kinematic mappings map displacements to points in a certain space. The main advantage is to work with points instead of displacements.

3.2.1 Study's kinematic Mapping

In section 1.2.2 we introduced the kinematic mapping of Study. We recall Study's mapping that maps displacements to a sliced hyperquadric $S_2^6 \backslash V^3$

$$\zeta : \mathrm{SE}(3) \to S_2^6 \backslash V^3 \subseteq \mathbb{P}^7(\mathbb{R}), \tag{3.4}$$
$$\mathrm{SE}(3) \ni \alpha \mapsto A = (a_0, a_1, a_2, a_3, c_0, c_1, c_2, c_3)^{\mathrm{T}}.$$

So far we did not mention that the exceptional generator V^3 is the vertex of a singular quadric

$$N_2^6 : a_0^2 + a_1^2 + a_2^2 + a_3^2 = 0.$$

The group of collineations in the image space that corresponds to SE(3) are the collineations $\mathrm{PGL}(\mathbb{P}^7, [S_2^6, N_2^6])$ that leave the pencil of quadrics $[S_2^6, N_2^6]$ invariant. Note that the image space is not a Cayley-Klein space. On one hand we could interpret the seven-dimensional projective space with its absolute figure

$$\mathcal{F} : S_2^6 \supset V^3 \supset Q_{4,0}^2$$

as Cayley-Klein space when we use the real numbers as underlying field. On the other hand, if we look with complex glasses on the scene, the exceptional generator V^3 becomes the vertex space of a singular quadric $N_2^6 : \sum_{i=0}^{3} a_i^2 = 0$, that belongs to the absolute figure. Furthermore, the quadric $Q_{4,0}^2$ is a quadric in the subspace V^3 given by the equation $c_0^2 + c_1^2 + c_2^2 + c_3^2 = 0$.

3.2.2 A Mapping for planar Displacements

Here, we present another kinematic mapping introduced by W. BLASCHKE [6] and J. GRÜNWALD [26]. The interested reader is referred to [13, Ch. 11] for further information. Planar Euclidean displacements can be decomposed into a rotation and a translation

$$\begin{pmatrix} x \\ y \end{pmatrix} \mapsto \begin{pmatrix} x' \\ y' \end{pmatrix} = \begin{pmatrix} \cos\varphi & -\sin\varphi \\ \sin\varphi & \cos\varphi \end{pmatrix} \begin{pmatrix} x \\ y \end{pmatrix} + \begin{pmatrix} a \\ b \end{pmatrix}.$$

The kinematic mapping $\varkappa$ maps planar Euclidean displacements to points of $\mathbb{P}^3(\mathbb{R})$. Note that the homogeneous coordinates of points are related to Cartesian coordinates $(x, y, z)^\mathrm{T}$ via

$$x = \frac{x_1}{x_0}, \quad y = \frac{x_2}{x_0}, \quad z = \frac{x_3}{x_0}, \quad \text{if } x_0 \neq 0.$$

Again not every point of the image space corresponds to a displacement. In analogy to section 3.2.1 the line $\ell: x_0 = x_3 = 0$ is the real vertex space of a reducible quadric $N_2^2 : x_0^2 + x_3^2 = 0$, which consists of the complex conjugate planes $x_0 + ix_3 = 0$ and $x_0 - ix_3 = 0$. Therefore, ℓ is the line of intersection of these two planes. To get a bijection we have to remove the line at infinity ℓ from the image space. Under these conditions, the map

$$\varkappa : \mathrm{SE}(2) \to \mathbb{P}(\mathbb{R})\setminus\ell \tag{3.5}$$

$$\mathrm{SE}(2) \ni \alpha(\varphi, a, b) \mapsto \left(2\cos\frac{\varphi}{2}, a\sin\frac{\varphi}{2} - b\cos\frac{\varphi}{2}, a\cos\frac{\varphi}{2} + b\sin\frac{\varphi}{2}, 2\sin\frac{\varphi}{2}\right)^\mathrm{T}$$

is one-to-one and onto. Hence, the image space is a quasi-elliptic space, cf. [25] with absolute figure

$$\mathcal{F} : Q_{2,0}^2 \supset A^1 \supset Q_{2,0}^0,$$

$Q_{2,0}^2 : x_0^2 + x_3^2 = 0, \quad A^1 : x_0 = x_3 = 0, \quad Q_{2,0}^0 : x_1^2 + x_2^2 = 0.$

Planar Euclidean displacements can be described by a subgroup of the group $\mathbb{U}_d$, see section 1.1.2. For example: If we want to describe planar Euclidean displacements in the $[x, y]$-plane, we have to restrict the group $\mathbb{U}_d$ to rotations around axes that are parallel to the z-axis and to translations in the $[x, y]$-plane.

3.3 Kinematic Mappings via Clifford Algebras

Both examples presented in the previous section can be treated together in the unifying and more general framework of Clifford algebras. In the following we shall explain how this works and illustrate the construction at hand of the previous examples. The Pin group corresponding to a certain homogeneous Clifford algebra model for a Cayley-Klein space is a double cover of the group of collineations that preserve the absolute figure $\mathcal{F}$. Moreover, the Spin group forms a double cover of the group of direct isometries of the corresponding Cayley-Klein space. Therefore, we just have to look at the Spin group as a subset of a projective space of the dimension 2^{n-1}.

3.3.1 Study's Mapping via Clifford Algebra

Again we start with the homogeneous model for the three-dimensional Euclidean space $\mathcal{Cl}_{(3,0,1)}$. The Spin group is located in the even part $\mathcal{Cl}^+_{(3,0,1)}$ of the algebra and every Spin group element satisfies the condition $N(\mathfrak{g}) = \mathfrak{g}\mathfrak{g}^* = 1$.

Remark 3.2. *Note, that for $n < 6$ the elements $\mathfrak{g}$ of the Spin or Pin group are characterized by $\mathfrak{g} = \pm 1$. For higher dimensions the sandwich action of an element with $\mathfrak{g}\mathfrak{g}^* = 1$ does not need to result in a vector when it is applied to a vector, see [21, p. 30].*

Thus, we take an arbitrary element of the even part of the algebra

$$\mathfrak{g} = a_0 e_0 + a_3 e_{12} + a_2 e_{13} + c_1 e_{14} + a_1 e_{23} + c_2 e_{24} + c_3 e_{34} + c_0 e_{1234}$$

and calculate the product with its conjugate element

$$\mathfrak{g}\mathfrak{g}^* = (a_0^2 + a_1^2 + a_2^2 + a_3^2)e_0 + 2(c_0 a_0 - a_1 c_1 + c_2 a_2 - c_3 a_3)e_{1234} = 1. \quad (3.6)$$

Eq. (3.6) shows that $\mathfrak{g} \in \mathrm{Spin}_{(3,0,1)}$ if the quadratic equation in the pseudoscalar part is equal to zero. With the isomorphism (3.3) we can see that this condition is exactly the equation of Study's quadric. The values $a_i, c_i, i = 0, \ldots, 3$ are the coordinates of a Euclidean displacement, for

$$c_0 a_0 - a_1 c_1 + c_2 a_2 - c_3 a_3 = 0. \quad (3.7)$$

Furthermore, the scalar part of Eq. (3.6) has to be equal to 1. Now we identify the eight components $a_0, \ldots, a_3, c_0, \ldots, c_3$ with coordinates of a seven-dimensional projective space $\mathbb{P}^7(\mathbb{R})$. The condition in the scalar part can now be relaxed to

$$a_0^2 + a_1^2 + a_2^2 + a_3^2 \neq 0.$$

This equation can be identified as the condition that a point is not allowed to lie in the exceptional space V^3. Hence, we have found a mapping from $\mathrm{Spin}_{(3,0,1)} \to S_2^6 \backslash V^3$, compare to Eq. (3.4). Note that $\mathfrak{g}$ and $-\mathfrak{g}$ represent the same displacement.

Matrices of Displacements In this paragraph we show how to describe the collineations of $\mathbb{P}^3(\mathbb{R})$ that preserve the absolute figure (3.2) of the Euclidean space seen as Cayley-Klein space and correspond to direct isometries. To do so, we look at the action of the Spin group on a point $P = (x_0, x_1, x_2, x_3)^\mathrm{T}$ that is represented by a grade-3 element

$$\mathfrak{p} = x_0 e_{123} + x_1 e_{234} + x_2 e_{134} + x_3 e_{124}.$$

An arbitrary element of the Spin group is given by

$$\mathfrak{g} = a_0 e_0 + a_1 e_{23} + a_2 e_{13} + a_3 e_{12} + c_0 e_{1234} + c_1 e_{14} + c_2 e_{24} + c_3 e_{34}, \quad \mathfrak{g}\mathfrak{g}^* = 1.$$

In the projective representation the condition $\mathfrak{g}\mathfrak{g}^* = 1$ changes to $\sum_{i=0}^{3} a_i^2 \neq 0$ and further we have $a_0 c_0 - a_1 c_1 + a_2 c_2 - a_3 c_3 = 0$. The action of a Spin group element can be written in terms of the sandwich operator

$$\begin{aligned}
\mathfrak{g}\mathfrak{p}\mathfrak{g}^* = &\ (a_0^2 + a_1^2 + a_2^2 + a_3^2) x_0 e_{123} \\
&+ \bigl(2(-a_0 c_1 - a_1 c_0 - a_2 c_3 - a_3 c_2) x_0 + (a_0^2 + a_1^2 - a_2^2 - a_3^2) x_1 \\
&+ 2(a_2 a_1 - a_0 a_3) x_2 + 2(a_0 a_2 + a_3 a_1) x_3 \bigr) e_{234} \\
&+ \bigl(2(a_0 c_2 + a_1 c_3 - a_2 c_0 - a_3 c_1) x_0 + 2(a_0 a_3 + a_2 a_1) x_1 \\
&+ (a_0^2 - a_1^2 + a_2^2 - a_3^2) x_2 + 2(a_3 a_2 - a_0 a_1) x_3 \bigr) e_{134} \\
&+ \bigl((2 a_2 c_1 + 2 a_1 c_2 - 2 a_0 c_3 - 2 a_3 c_0) x_0 + 2(a_3 a_1 - a_0 a_2) x_1 \\
&+ 2(a_3 a_2 + a_0 a_1) x_2 + (a_0^2 - a_1^2 - a_2^2 + a_3^2) x_3 \bigr) e_{124}.
\end{aligned}$$

3.3 Kinematic Mappings via Clifford Algebras

Here, we see that the action of $\mathfrak{g}$ on $\mathfrak{p}$ is linear. Hence, we can write this action as a product of a matrix with a vector

$$x' = \mathrm{M} \cdot x, \quad x = (x_0, x_1, x_2, x_3)^{\mathrm{T}}, \quad x' = (x'_0, x'_1, x'_2, x'_3)^{\mathrm{T}}$$

with

$$\mathrm{M} = \frac{1}{\Delta}\begin{pmatrix} \Delta & 0 & 0 & 0 \\ l & a_0^2 + a_1^2 - a_2^2 - a_3^2 & 2(a_2a_1 - a_0a_3) & 2(a_0a_2 + a_3a_1) \\ m & 2(a_0a_3 + a_2a_1) & a_0^2 - a_1^2 + a_2^2 - a_3^2 & 2(a_3a_2 - a_0a_1) \\ n & 2(a_3a_1 - a_0a_2) & 2(a_3a_2 + a_0a_1) & a_0^2 - a_1^2 - a_2^2 + a_3^2 \end{pmatrix}$$

and

$$\Delta = a_0^2 + a_1^2 + a_2^2 + a_3^2, \qquad l = 2(-a_0c_1 - a_3c_2 - a_2c_3 - a_1c_0),$$
$$m = 2(a_0c_2 + a_1c_3 - a_2c_0 - a_3c_1), \quad n = 2(a_2c_1 + a_1c_2 - a_0c_3 - a_3c_0).$$

The factor $\frac{1}{\Delta}$ guarantees that the right lower 3×3 submatrix is an element of $\mathrm{SO}(3)$.

Collineations in the Image Space Transformations in the underlying geometry induce collineations in the kinematic image space. Therefore, we are interested in the matrix representation of Spin group elements, see Eq. (1.34). Thus, we take an element from $\mathrm{Spin}_{(3,0,1)}$

$$\mathfrak{g} = a_0e_0 + a_1e_{23} + a_2e_{13} + a_3e_{12} + c_0e_{1234} + c_1e_{14} + c_2e_{24} + c_3e_{34}.$$

When using homogeneous coordinates the norming condition becomes $\sum_{i=0}^{3} a_i^2 \neq 0$. Furthermore, the parameters $(a_0, \ldots, a_3, c_0, \ldots, c_3)$ have to satisfy the condition (3.7). We get

$$\mathfrak{g}e_0 = a_0e_0 + a_3e_{12} + a_2e_{13} + c_1e_{14} + a_1e_{23} + c_2e_{24} + c_3e_{34} + c_0e_{1234},$$
$$\mathfrak{g}e_{23} = -a_1e_0 - a_2e_{12} + a_3e_{13} - c_0e_{14} + a_0e_{23} - c_3e_{24} + c_2e_{34} + c_1e_{1234},$$
$$\mathfrak{g}e_{13} = -a_2e_0 + a_1e_{12} + a_0e_{13} - c_3e_{14} - a_3e_{23} + c_0e_{24} + c_1e_{34} - c_2e_{1234},$$
$$\mathfrak{g}e_{12} = -a_3e_0 + a_0e_{12} - a_1e_{13} - c_2e_{14} + a_2e_{23} + c_1e_{24} - c_0e_{34} + c_3e_{1234},$$
$$\mathfrak{g}e_{1234} = -a_1e_{14} + a_2e_{24} - a_3e_{34} + a_0e_{1234},$$
$$\mathfrak{g}e_{14} = a_0e_{14} - a_3e_{24} - a_2e_{34} + a_1e_{1234},$$

$$\mathfrak{g}e_{24} = a_3 e_{14} + a_0 e_{24} - a_1 e_{34} - a_2 e_{1234},$$
$$\mathfrak{g}e_{34} = a_2 e_{14} + a_1 e_{24} + a_0 e_{34} + a_3 e_{1234}.$$

If we write these equations in matrix form, we obtain a representation of collineations of $\mathbb{P}^7(\mathbb{R})$ that preserve the pencil of quadrics spanned by $[S_2^6, N_2^6]$:

$$[G^+] = \begin{pmatrix} a_0 & -a_1 & -a_2 & -a_3 & 0 & 0 & 0 & 0 \\ a_1 & a_0 & -a_3 & a_2 & 0 & 0 & 0 & 0 \\ a_2 & a_3 & a_0 & -a_1 & 0 & 0 & 0 & 0 \\ a_3 & -a_2 & a_1 & a_0 & 0 & 0 & 0 & 0 \\ c_0 & c_1 & -c_2 & c_3 & a_0 & a_1 & -a_2 & a_3 \\ c_1 & -c_0 & -c_3 & -c_2 & -a_1 & a_0 & a_3 & a_2 \\ c_2 & -c_3 & c_0 & c_1 & a_2 & -a_3 & a_0 & a_1 \\ c_3 & c_2 & c_1 & -c_0 & -a_3 & -a_2 & -a_1 & a_0 \end{pmatrix}.$$

In order to write $[G^+]$ in the usual form, see Eq. (1.9) the isomorphism (3.3) has to be used.

3.3.2 Blaschke's and Grünwald's Mapping via Clifford Algebra

For planar Euclidean geometry the homogeneous Clifford algebra model $\mathcal{Cl}_{(2,0,1)}$ is adequate. We find the Spin group in the even part $\mathcal{Cl}^+_{(2,0,1)}$. A general element of the even part is given by

$$\mathfrak{g} = a_0 e_0 + a_1 e_{12} + c_0 e_{23} + c_1 e_{13}.$$

The condition that this element is a Spin group element reads now

$$\mathfrak{g}\mathfrak{g}^* = a_0^2 + a_1^2 = 1.$$

In this case we have just one equation in the scalar part. If we change to homogeneous coordinates, the condition gets $a_0^2 + a_1^2 \neq 0$. Again we have the equation of a pair of complex conjugate planes intersecting in a real line. This results in the same image space as in the previous section, see Eq. (3.5).

Matrices of planar Euclidean Displacements With the Spin group we can find the matrix representation corresponding to the group SE(2) as group of collineations fixing the absolute figure, cf. Eq. (3.1). All we have to do is to apply the sandwich operator to a point. As we choose a plane based construction a point $P = (x_0, x_1, x_2)^T$ can be described in the algebra $\mathcal{C}\ell_{(2,0,1)}$ via $\mathfrak{p} = x_0 e_{12} + x_1 e_{23} + x_2 e_{13}$. A general element of the Spin group is given by $\mathfrak{g} = a_0 e_0 + a_1 e_{12} + c_0 e_{23} + c_1 e_{13}$. Applying the sandwich operator to $\mathfrak{p}$ results in

$$\mathfrak{g}\mathfrak{p}\mathfrak{g}^* = (a_0^2 + a_1^2)x_0 e_{12} + (2(a_1 c_0 + a_0 c_1)x_0 + (a_0^2 - a_1^2)x_1 - 2a_0 a_1 x_2)e_{23}$$
$$+ (2(a_1 c_1 - a_0 c_0)x_0 + 2a_0 a_1 x_1 + (a_0^2 + a_1^2)x_2)e_{13}.$$

If we rewrite the action of this transformation as linear transformation of the projective plane, we get

$$\begin{pmatrix} x_0' \\ x_1' \\ x_2' \end{pmatrix} = \frac{1}{a_0^2 + a_1^2} \begin{pmatrix} a_0^2 + a_1^2 & 0 & 0 \\ 2(a_1 c_0 + a_0 c_1) & a_0^2 - a_1^2 & -2a_0 a_1 \\ 2(a_1 c_1 - c_0 a_0) & 2a_0 a_1 & a_0^2 - a_1^2 \end{pmatrix} \cdot \begin{pmatrix} x_0 \\ x_1 \\ x_2 \end{pmatrix}. \quad (3.8)$$

In the first line of Eq. (3.8) we see that the line at infinity is fixed under these collineations. It is clear that all planar Euclidean displacements are obtained in this way, because the Spin group is a double cover of SE(2). The factor is added artificially to guarantee that we just look at matrices belonging to the Spin group. These two examples show that the Spin group can be used to get the matrix group of the direct isometries of a Cayley-Klein geometry, if it is representable as homogeneous Clifford algebra model.

Collineations in the Image Space The Spin group elements induce in the image space a group of transformations that fix the pair of complex conjugate planes, their intersection line and the two complex conjugate points on this line. To get this group we have to look at the matrix representation of the Spin group. Let $\mathfrak{g} = a_0 e_0 + a_1 e_{12} + c_0 e_{23} + c_1 e_{13} \in \mathrm{Spin}_{(2,0,1)}$. Calculating the matrix representation as described in section 1.3.4 results in

$$\mathrm{PGL}(\mathbb{P}^2(\mathbb{R}), \mathcal{F}) = \left\{ \begin{pmatrix} a_0 & -a_1 & 0 & 0 \\ a_1 & a_0 & 0 & 0 \\ c_1 & -c_0 & a_0 & a_1 \\ c_0 & c_1 & -a_1 & a_0 \end{pmatrix} \middle| \, a_0, a_1, c_0, c_1 \in \mathbb{R} \right\}.$$

3.4 Kinematic mappings for other Cayley-Klein Spaces

This procedure can be applied to homogeneous Clifford algebra models of any signature. We will present one more example here. For the remaining signatures, see Table 3.1 and Table 3.2. Note that it makes no sense to look at the completely degenerate Clifford algebra $\mathcal{C}\ell_{(0,0,3)}$ because the scalar product of any two vectors is zero.

3.4.1 Two-dimensional Cayley-Klein Spaces

Elliptic Plane The homogeneous model for the elliptic plane is given by $\mathcal{C}\ell_{(3,0,0)}$. Let $\mathfrak{g} = a_0 e_0 + a_1 e_{12} + c_0 e_{23} + c_1 e_{13}$ be an element of the even part $\mathcal{C}\ell^+_{(3,0,0)}$. The product of this element with its conjugate element yields

$$\mathfrak{g}\mathfrak{g}^* = a_0^2 + a_1^2 + c_0^2 + c_1^2 = 1.$$

In homogeneous coordinates this condition gets $a_0^2 + a_1^2 + c_0^2 + c_1^2 \neq 0$. Therefore, we can interpret the kinematic image of the congruences in elliptic spaces as non-degenerate Cayley-Klein space with absolute figure $\mathcal{F} : Q_{4,0}^2$. Furthermore, this means, if we work over the real numbers, every point has a pre-image. In Table 3.1 the absolute

Table 3.1: Planar Cayley-Klein geometries and belonging Clifford algebras

CK space	absolute Figure $\mathcal{F}$	Clifford alg.
elliptic	$Q_{3,0}^1$	$\mathcal{C}\ell_{(3,0,0)}, \mathcal{C}\ell_{(0,3,0)}$
hyperbolic	$Q_{3,1}^1$	$\mathcal{C}\ell_{(2,1,0)}, \mathcal{C}\ell_{(1,2,0)}$
Euclidean	$Q_{1,0}^1 \supset A^1 \supset Q_{2,0}^0$	$\mathcal{C}\ell_{(2,0,1)}, \mathcal{C}\ell_{(0,2,1)}$
pseudo-Euclidean	$Q_{1,0}^1 \supset A^1 \supset Q_{2,1}^0$	$\mathcal{C}\ell_{(1,1,1)}$
quasi-elliptic	$Q_{2,0}^1 \supset A^0 \supset Q_{1,0}^{-1}$	$\mathcal{C}\ell_{(2,0,1)}, \mathcal{C}\ell_{(0,2,1)}$
quasi-hyperbolic	$Q_{2,1}^1 \supset A^0 \supset Q_{1,0}^{-1}$	$\mathcal{C}\ell_{(1,1,1)}$
totally isotropic	$Q_{1,0}^1 \supset A^1 \supset Q_{1,0}^0 \supset A^0 \supset Q_{1,0}^{-1}$	$\mathcal{C}\ell_{(1,0,2)}, \mathcal{C}\ell_{(0,1,2)}$

3.4 Kinematic mappings for other Cayley-Klein Spaces

figures $\mathcal{F}$ and the corresponding Clifford algebras for possible two-dimensional Cayley-Klein spaces are given. Note that the points of the Cayley-Klein geometry are always written in the grade-2 subspace of the Clifford algebra except the quasi-elliptic and the quasi-hyperbolic case. In these two cases we use the Poincaré isomorphism to describe these Cayley-Klein geometries as dual partners of the Euclidean respectively, the pseudo-Euclidean Cayley-Klein geometry. Therefore, points of the quasi-elliptic (quasi-hyperbolic) Cayley-Klein geometry are represented as grade-1 elements in the algebra corresponding to the Euclidean (pseudo-Euclidean) Cayley-Klein space. Table 3.2 shows the kinematic image spaces of the seven planar Cayley-Klein spaces. Note that every image space is again a Cayley-Klein space.

Table 3.2: Kinematic image spaces of two-dimensional Cayley-Klein geometries presented as three-dimensional Cayley-Klein spaces with their absolute figures $\mathcal{F}$

pre-image	CK image space	absolute Figure $\mathcal{F}$
elliptic	elliptic	$Q^2_{4,0}$
hyperbolic	hyperbolic idx. 1	$Q^2_{4,2}$
Euclidean	quasi-elliptic	$Q^2_{2,0} \supset A^1 \supset Q^0_{2,0}$
pseudo-Euclidean	quasi-hyperb. idx. 0	$Q^2_{2,1} \supset A^1 \supset Q^0_{2,1}$
quasi-elliptic	quasi-elliptic	$Q^2_{2,0} \supset A^1 \supset Q^0_{2,0}$
quasi-hyperbolic	quasi-hyperb. idx. 0	$Q^2_{2,1} \supset A^1 \supset Q^0_{2,1}$
totally isotropic	totally isotropic	$Q^2_{1,0} \supset A^2 \supset Q^1_{2,0} \supset A^0 \supset Q^{-1}_{1,0}$

3.4.2 Three-dimensional Cayley-Klein Spaces

The kinematic image space for the three-dimensional Euclidean space was constructed in section 3.3.1. Here we want to repeat this construction for the elliptic three-dimensional Cayley-Klein space. In [59, Ch. 11] the author showed that elements of the group SO(4) can be identified with points of Study's quadric. In this case no exceptions have to be made. Therefore, the kinematic mapping is one-to-one and onto. Let us look what happens, if we construct the kinematic mapping corresponding to the three-dimensional elliptic Cayley-Klein geometry which is modelled through $\mathcal{Cl}_{(4,0,0)}$. This Clifford algebra is of dimension 16 and the even part $\mathcal{Cl}^+_{(4,0,0)}$ is eight-dimensional. An arbitrary Spin group element is given by

$$\mathfrak{g} = a_0 e_0 + a_1 e_{23} + a_2 e_{13} + a_3 e_{12} + c_0 e_{1234} + c_1 e_{14} + c_2 e_{24} + c_3 e_{34}.$$

The condition that it is an element of the Spin group reads

$$\mathfrak{g}\mathfrak{g}^* = (a_0^2 + a_1^2 + a_2^2 + a_3^2 + c_0^2 + c_1^2 + c_2^2 + c_3^2)e_0 + (a_0 c_0 - a_1 c_1 + a_2 c_2 - a_3 c_3)e_{1234} = 1.$$

If we change to seven-dimensional projective space, the first condition is that the scalar part should not vanish

$$a_0^2 + a_1^2 + a_2^2 + a_3^2 + c_0^2 + c_1^2 + c_2^2 + c_3^2 \neq 0.$$

As in the case for Euclidean displacements this defines the exceptional set, i.e., points that do not have a pre-image. Here, we can write this exceptional set as hyperquadric

$$Q_2^6 : a_0^2 + a_1^2 + a_2^2 + a_3^2 + c_0^2 + c_1^2 + c_2^2 + c_3^2 = 0.$$

The term in the pseudoscalar part is again the equation of Study's quadric

$$S_2^6 : a_0 c_0 - a_1 c_1 + a_2 c_2 - a_3 c_3 = 0.$$

Every point on Study's quadric stands for an element of SO(4) since Spin$_{(4,0,0)}$ is a double cover of SO(4). In this case it is not necessary to slice the quadric, since Q_2^6 has no real point. Furthermore, this means the pencil of quadrics spanned by $[S_2^6, Q_2^6]$ is the absolute figure of the kinematic image space. Surprisingly, every kinematic mapping of three-dimensional Cayley-Klein spaces maps the Spin group elements to points on Study's quadric. In the following we list the possible Cayley-Klein spaces and the absolute quadric pencil in the kinematic image space. In Table 3.3 the other possible homogeneous Clifford algebra models are presented. Note that every Cayley-Klein space is self-dual except the Euclidean and the pseudo-Euclidean. Their dual Cayley-Klein spaces can be represented in the dual homogeneous Clifford algebra model obtained by the Poincaré duality.

3.4 Kinematic mappings for other Cayley-Klein Spaces

Table 3.3: Three-dimensional Cayley-Klein spaces with possible Clifford algebra representation and exceptional quadric Q_2^6

CK-space	$\mathcal{Cl}_{(p,q,r)}$	Q_2^6
elliptic	$\mathcal{Cl}_{(4,0,0)}, \mathcal{Cl}_{(0,4,0)}$	$\sum_{i=0}^{3} a_i^2 + \sum_{i=0}^{3} c_i^2$
hyperbolic idx. 0	$\mathcal{Cl}_{(3,1,0)}, \mathcal{Cl}_{(1,3,0)}$	$\sum_{i=0}^{3} a_i^2 - \sum_{i=0}^{3} c_i^2$
hyperbolic idx. 1	$\mathcal{Cl}_{(2,2,0)}$	$a_0^2 - a_1^2 - a_2^2 + a_3^2 + c_0^2 - c_1^2 - c_2^2 + c_3^2$
Euclidean	$\mathcal{Cl}_{(3,0,1)}, \mathcal{Cl}_{(0,3,1)}$	$\sum_{i=0}^{3} a_i^2$
pseudo-Euclidean	$\mathcal{Cl}_{(2,1,1)}, \mathcal{Cl}_{(1,2,1)}$	$a_0^2 - a_1^2 - a_2^2 + a_3^2$
quasi-elliptic	$\mathcal{Cl}_{(2,0,2)}, \mathcal{Cl}_{(0,2,2)}$	$a_0^2 + a_3^2$
quasi-hyperb. idx. 0	$\mathcal{Cl}_{(1,1,2)}$	$a_0^2 - a_3^2$
double isotropic	$\mathcal{Cl}_{(1,0,3)}, \mathcal{Cl}_{(0,1,3)}$	a_0^2

3.4.3 Higher dimensional kinematic Mappings

The procedure presented above can be generalized to arbitrary dimensions. Here, we present just one example, the four-dimensional Euclidean space. Written as homogeneous Clifford algebra model we get $\mathcal{Cl}_{(4,0,1)}$. The dimension of the algebra is 32, and therefore, the dimension of the even part is 16. Now we take an arbitrary element of the even part and formulate the conditions that it is element of $\mathrm{Spin}_{(4,0,1)}$. Note that in this case two conditions are necessary, i.e., $\mathfrak{g}\mathfrak{g}^* = \mathfrak{g}^*\mathfrak{g} = 1$. The element has the form

$$\mathfrak{g} = a_0 e_0 + a_1 e_{23} + a_2 e_{13} + a_3 e_{12} + a_4 e_{15} + a_5 e_{45} + a_6 e_{25} + a_7 e_{35}$$
$$+ c_1 e_{14} + c_2 e_{24} + c_3 e_{34} + c_4 e_{1235} + c_5 e_{1345} + c_6 e_{1245} + c_7 e_{2345} + c_0 e_{1234}.$$

The product $\mathfrak{g}\mathfrak{g}^*$ calculates to

$$\mathfrak{g}\mathfrak{g}^* = (c_2^2 + c_3^2 + a_2^2 + a_3^2 + a_0^2 + c_1^2 + a_1^2 + c_0^2) e_0 \qquad (3.9)$$
$$+ 2(c_0 a_0 - a_1 c_1 - c_3 a_3 + c_2 a_2) e_{1234}$$
$$+ 2(c_7 c_1 + c_4 a_0 - c_5 c_2 - c_0 a_5 + a_6 a_2 - a_1 a_4 + c_6 c_3 - a_7 a_3) e_{1235}$$
$$+ 2(c_5 a_1 + a_6 c_1 - a_5 a_3 - c_2 a_4 - c_7 a_2 + c_6 a_0 + c_0 a_7 - c_4 c_3) e_{1245}$$
$$+ 2(c_4 c_2 - c_6 a_1 - a_5 a_2 - c_3 a_4 - c_0 a_6 + c_5 a_0 + c_7 a_3 + a_7 c_1) e_{1345}$$

$$+ 2(c_6 a_2 + c_0 a_4 - c_4 c_1 - c_5 a_3 + a_7 c_2 - c_3 a_6 + c_7 a_0 - a_5 a_1)e_{2345}.$$

Furthermore, we have to calculate $\mathfrak{g}^*\mathfrak{g}$

$$\mathfrak{g}^*\mathfrak{g} = (a_1^2 + c_2^2 + c_3^2 + c_0^2 + c_1^2 + a_3^2 + a_2^2 + a_0^2)e_0 \qquad (3.10)$$
$$+ 2(c_2 a_2 - c_3 a_3 + c_0 a_0 - a_1 c_1)e_{1234}$$
$$+ 2(c_4 a_0 - a_7 a_3 - c_7 c_1 + c_0 a_5 - a_1 a_4 + a_6 a_2 - c_6 c_3 + c_5 c_2)e_{1235}$$
$$+ 2(c_7 a_2 + a_6 c_1 - c_5 a_1 - a_5 a_3 + c_6 a_0 - c_2 a_4 + c_4 c_3 - c_0 a_7)e_{1245}$$
$$+ 2(c_0 a_6 - c_3 a_4 - c_4 c_2 + c_6 a_1 - c_7 a_3 + a_7 c_1 + c_5 a_0 - a_5 a_2)e_{1345}$$
$$+ 2(a_7 c_2 + c_4 c_1 - a_5 a_1 + c_5 a_3 - c_6 a_2 - c_3 a_6 - c_0 a_4 + c_7 a_0)e_{2345}.$$

Vanishing coefficients of the grade-4 generators in Eq. (3.9) and (3.10) are the conditions for an even grade element to lay in the Spin group. All coefficients of grade-4 elements have to vanish and the scalar part must be different from zero. Some quadric equations occur in both expressions. All in all, the kinematic image space is a projective variety in $\mathbb{P}^{15}(\mathbb{R})$ sliced along a quadric. It can be written as the intersection of nine quadrics minus one quadric and therefore, as pseudo algebraic variety:

$$\mathcal{V}: \bigcap_{i=1}^{9} Q_i \setminus N_1 \subseteq \mathbb{P}^{15}(\mathbb{R}),$$

with

$N_1 : c_2^2 + c_3^2 + a_2^2 + a_3^2 + a_0^2 + c_1^2 + a_1^2 + c_0^2 = 0,$
$Q_1 : c_0 a_0 - a_1 c_1 - c_3 a_3 + c_2 a_2 = 0,$
$Q_2 : c_7 c_1 + c_4 a_0 - c_5 c_2 - c_0 a_5 + a_6 a_2 - a_1 a_4 + c_6 c_3 - a_7 a_3 = 0,$
$Q_3 : c_5 a_1 + a_6 c_1 - a_5 a_3 - c_2 a_4 - c_7 a_2 + c_6 a_0 + c_0 a_7 - c_4 c_3 = 0,$
$Q_4 : c_4 c_2 - c_6 a_1 - a_5 a_2 - c_3 a_4 - c_0 a_6 + c_5 a_0 + c_7 a_3 + a_7 c_1 = 0,$
$Q_5 : c_6 a_2 + c_0 a_4 - c_4 c_1 - c_5 a_3 + a_7 c_2 - c_3 a_6 + c_7 a_0 - a_5 a_1 = 0,$
$Q_6 : c_4 a_0 - a_7 a_3 - c_7 c_1 + c_0 a_5 - a_1 a_4 + a_6 a_2 - c_6 c_3 + c_5 c_2 = 0,$
$Q_7 : c_7 a_2 + a_6 c_1 - c_5 a_1 - a_5 a_3 + c_6 a_0 - c_2 a_4 + c_4 c_3 - c_0 a_7 = 0,$
$Q_8 : c_0 a_6 - c_3 a_4 - c_4 c_2 + c_6 a_1 - c_7 a_3 + a_7 c_1 + c_5 a_0 - a_5 a_2 = 0,$
$Q_9 : a_7 c_2 + c_4 c_1 - a_5 a_1 + c_5 a_3 - c_6 a_2 - c_3 a_6 - c_0 a_4 + c_7 a_0 = 0.$

Remark 3.3. *With this method it is also possible to construct the mapping* $SO(3) \to S^3$. *In this case we do not need the homogeneous Clifford algebra model. Therefore, we take the affine Clifford algebra model* $C\ell_{(3,0,0)}$ *and do the same construction for a Spin group element.*

3.5 Projective Varieties via kinematic Algebra Elements

Here, we present another method to construct projective varieties corresponding to the Spin group via Clifford algebra. Therefore, a definition is needed.

Definition 3.2. *An element* $\mathfrak{g}$ *of a Clifford algebra* $C\ell_{(p,q,r)}$ *is called* kinematic, *if it satisfies the following equation*

$$\mathfrak{g}^2 = tr(\mathfrak{g})\mathfrak{g} - N(\mathfrak{g}),$$

where $tr(\mathfrak{g}) = \mathfrak{g} + \mathfrak{g}^*$ *is the trace of the element.*

Def. 3.2 generalizes the definition of kinematic algebras over fields, see [39]. To obtain projective varieties by using this definition, we first show that every Spin group element is kinematic.

Lemma 3.1. *Spin group elements* $\mathfrak{g} \in C\ell_{(p,q,r)}$ *are kinematic, i.e., they fulfil the equation*

$$\mathfrak{g}^2 = tr(\mathfrak{g})\mathfrak{g} - N(\mathfrak{g}).$$

Proof. The proof is done by direct calculation.

$$\mathfrak{g}^2 = tr(\mathfrak{g})\mathfrak{g} - N(\mathfrak{g}) = (\mathfrak{g} + \mathfrak{g}^*)\mathfrak{g} - \mathfrak{g}\mathfrak{g}^* = \mathfrak{g}\mathfrak{g} + \mathfrak{g}^*\mathfrak{g} - \mathfrak{g}\mathfrak{g}^* = \mathfrak{g}\mathfrak{g}.$$

Here we have used, that $\mathfrak{g}\mathfrak{g}^* = \mathfrak{g}^*\mathfrak{g} = 1$ for every Spin group element. □

Now we ask for general conditions that have to be satisfied by kinematic elements. We aim at a projective variety that corresponds to the Spin group. Thus, we take an arbitrary element $\mathfrak{g} \in C\ell_{(p,q,r)}^+$ and calculate the conditions from

$$\mathfrak{g}\mathfrak{g}^* - tr(\mathfrak{g})\mathfrak{g} - N(\mathfrak{g}) = 0.$$

This results in quadratic equations in several generators that all have to vanish. Now we use these equations and check the condition $\mathfrak{g}\mathfrak{g}^* \neq 0$ to get more conditions that will guarantee that the element is in the Spin group. To understand this method we present the kinematic mapping for the group of four-dimensional Euclidean displacements, i.e., $\mathrm{Spin}_{(4,0,1)}$.

Example 3.1. *A general element* $\mathfrak{g} \in \mathcal{C}\ell^+_{(4,0,1)}$ *has the form*

$$\mathfrak{g} = a_0 e_0 + a_1 e_{23} + a_2 e_{13} + a_3 e_{12} + a_4 e_{15} + a_5 e_{45} + a_6 e_{25} + a_7 e_{35}$$
$$+ c_1 e_{14} + c_2 e_{24} + c_3 e_{24} + c_4 e_{1235} + c_5 e_{1345} + c_6 e_{1245} + c_7 e_{2345} + c_0 e_{1234}.$$

For kinematic elements we have

$$0 = \mathfrak{g}\mathfrak{g}^* - tr(\mathfrak{g}) - N(\mathfrak{g}).$$

In expanded form this condition reads

$$0 = 4(c_6 a_2 + c_0 a_4 - c_4 c_1 - c_5 a_3)e_{2345} - 4(-c_7 c_1 - c_6 c_3 + c_5 c_2 + c_0 a_5)e_{1235}$$
$$+ 4(-c_7 a_2 + c_0 a_7 + c_5 a_1 - c_4 c_3)e_{1245} - 4(c_0 a_6 + c_6 a_1 - c_4 c_2 - c_7 a_3)e_{1345}.$$

Now we use these four quadratic equations and formulate the Spin group condition $\mathfrak{g}\mathfrak{g}^* = 1$,

$$\begin{aligned}\mathfrak{g}\mathfrak{g}^* &= (a_1^2 + c_2^2 + c_3^2 + c_0^2 + c_1^2 + a_0^2 + a_3^2 + a_2^2)e_0 \\ &+ 2(-a_1 c_1 - c_3 a_3 + c_0 a_0 + c_2 a_2)e_{1234} \\ &+ 2(-c_3 a_6 + a_7 c_2 + c_7 a_0 - a_5 a_1)e_{2345} \\ &+ 2(c_4 a_0 - a_7 a_3 + a_6 a_2 - a_1 a_4)e_{1235} \\ &+ 2(c_6 a_0 - a_5 a_3 + a_6 c_1 - c_2 a_4)e_{1245} \\ &+ 2(a_7 c_1 - a_5 a_2 - c_3 a_4 + c_5 a_0)e_{1345}\end{aligned}$$

and get five more quadratic equations that have to vanish. All in all we have the corresponding projective variety described by nine quadratic equations and one exceptional quadric with equation

$$N_1 : a_0^2 + a_1^2 + a_2^2 + a_3^2 + c_0^2 + c_1^2 + c_2^2 + c_3^2 = 0,$$
$$R_1 : c_6 a_2 + c_0 a_4 - c_4 c_1 - c_5 a_3 = 0,$$
$$R_2 : -c_7 c_1 - c_6 c_3 + c_5 c_2 + c_0 a_5 = 0,$$

3.5 Projective Varieties via kinematic Algebra Elements

$$R_3 : -c_7a_2 + c_0a_7 + c_5a_1 - c_4c_3 = 0,$$
$$R_4 : c_0a_6 + c_6a_1 - c_4c_2 - c_7a_3 = 0,$$
$$R_5 : -a_1c_1 - c_3a_3 + c_0a_0 + c_2a_2 = 0,$$
$$R_6 : -c_3a_6 + a_7c_2 + c_7a_0 - a_5a_1 = 0,$$
$$R_7 : c_4a_0 - a_7a_3 + a_6a_2 - a_1a_4 = 0,$$
$$R_8 : c_6a_0 - a_5a_3 + a_6c_1 - c_2a_4 = 0,$$
$$R_9 : a_7c_1 - a_5a_2 - c_3a_4 + c_5a_0 = 0.$$

This method results in an ideal that describes the same projective variety

$$\mathcal{V} : \bigcap_{i=1}^{9} R_i \backslash N_1 \subseteq \mathbb{P}^{15}(\mathbb{R}),$$

as in the previous section. Therefore, we give the R_i as linear combinations of the Q_i

$$R_1 = \frac{1}{2}(Q_9 - Q_5), \quad R_2 = \frac{1}{2}(Q_6 - Q_2), \quad R_3 = \frac{1}{2}(Q_3 - Q_7),$$
$$R_4 = \frac{1}{2}(Q_8 - Q_4), \quad R_5 = Q_1, \quad R_6 = \frac{1}{2}(Q_9 + Q_5),$$
$$R_7 = \frac{1}{2}(Q_6 + Q_2), \quad R_8 = \frac{1}{2}(Q_3 + Q_7), \quad R_9 = \frac{1}{2}(Q_8 + Q_4).$$

The advantage of both presented methods is that we can calculate a Gröbner basis for the ideal and apply the theory of ideals to the constructed point models.

Conclusion

In chapter 1 we presented geometric models and representation that are used today. Afterwards, Clifford algebras, the homogeneous Clifford algebra model, and the conformal model were introduced. Moreover, we presented a method to transfer general projective transformations acting in three-dimensional projective space to elements of the homogeneous Clifford algebra model $\mathcal{C}\ell_{(3,3,0)}$. All entities known from line geometry occur naturally in this model and can be transformed projectively by the application of the sandwich operator. It was shown that the sandwich action of non-null vectors corresponds to regular null polarities, *i.e.*, correlations that are involutions as the basic elements building up the group of regular projective transformations. The sandwich action of null vectors corresponds to singular null polarities and we proved that every regular projective transformation can be expressed as the product of six null polarities at the most. Furthermore, we present a Clifford algebra model that serves as a generalization of the conformal geometric algebra. A generalization of inversions with respect to conics, quadrics, and even hyperquadrics in any dimension is possible with the use of the sandwich operator. Hyperquadrics in principal position are simply represented as grade-1 elements. Classical representations of groups are embedded in this algebra naturally.

The basics of chain geometry were introduced in chapter 2. This theory was applied to Clifford algebras. We showed how the cross ratio can be used to parametrize chains. Moreover, it was proven that the connected components of the Pin- and Spin groups define subspaces of chain spaces. Therefore, every element contained in a chain that is defined by three elements of the same connected component of the Pin- or the Spin group is contained in the same connected component of the Pin- or the Spin group. Of special interest was the question for the cross ratio of dual unit quaternions that was answered in detail.

Moreover, we derived an algebraic biarc construction with contact spaces. Chains of the grade-1 subspace that are in contact at a certain point are parametrized with the cross ratio and correspond to conic sections.

Old and well-known kinematic mappings were unified in chapter 3 in one framework by the use of Clifford algebras. Collineations in any kinematic image and the corresponding Cayley-Klein space can be derived from the homogeneous Clifford algebra model. We presented the Euclidean spaces of dimension two and three in detail as example. Furthermore, a general method to construct pseudo algebraic varieties in certain projective spaces was presented. These pseudo algebraic varieties are point models for the Spin group of the homogeneous Clifford algebra model respectively the collineation group of a certain Cayley-Klein space. We performed the construction for the four-dimensional Euclidean. Moreover, the general construction enables the examination of new kinematic image spaces for possible Pin and Spin groups. This construction can also be done for Pin groups or for any other subgroup of a Clifford group. Due to the fact that projective varieties correspond to ideals, methods of Gröbner basis calculus can now be applied to kinematic image spaces.

Index

absolute figure, 182
admissible pairs, 103
algebra, 102
    Clifford, 26
    degenerate, 26
    geometric, 26

Benz planes, 116
biarc, 165
Blaschke cylinder, 117

Cayley-Klein
    geometry, 182
    space, 182, 193
center, 102
chain, 106
    geometry, 106
    space, 107
Clifford algebra
    conjugation, 29
    even part, 28
    main involution, 30
    odd part, 28
    universal, 28
Clifford group, 33
contact
    relation, 121
    space, 121
cross ratio, 109

distance
    relation, 101
    space, 101
dual
    number, 8
    orthogonal matrix, 13
    quaternion, 10
    Rodriguez formula, 14
    sphere, 136
    vector, 9

general linear group, 102
geometric inner product null space, 44
geometric outer product null space, 44
Grassmann
    algebra, 39
    coordinates, 39, 40
group
    Spin group, 35

horosphere, 43

ideal, 199
inner product null space, 32

Jordan-closed, 119
Jordan-system, 118, 119

kinematic algebra element, 197
kinematic mapping, 21, 185, 192
    of Blaschke and Grünwald, 185
    of Study, 185
Klein's quadric, 131

Laguerre geometry, 109
Lie's quadric, 73
Lorentz transformation, 48

Möbius geometry, 108
Minkowski geometry, 109

outer product null space, 31

parallel, 102
Pin group, 34
pitch, 20
Plücker coordinates, 15
Poincaré duality, 41
projective linear group, 104
projective variety, 199
pseudoscalar, 46

quadratic form, 114
quadric
    Klein's quadric, 15, 50, 176
    Lie's quadric, 133
    Study's quadric, 21, 76, 135
quaternion, 6
    dual, 10
    unit, 7

sandwich operator, 7
simple blade, 31
standard basis, 28
strong Jordan-system, 119
Study's sphere, 25

subalgebra, 118

trace, 197
transitive
    2-Δ-transitive, 104
    3-Δ-transitive, 104

versor, 28

List of Symbols

$(\mathcal{P}, \Delta)$	distance space, p. 102
(l, m)	Plücker coordinates with direction vector l and moment vector m, p. 17
(p, q, r)	signature of a Clifford algebra, p. 28
α	main involution, p. 30
$\bigwedge V$	exterior algebra or Grassmann algebra, p. 39
χ	bijection between quadrics in principal position and vectors in QnGA, p. 82
$\mathcal{C}\ell_{(p,q,r)}^+$	even part of $\mathcal{C}\ell_{(p,q,r)}$, p. 28
$\mathcal{C}\ell_{(p,q,r)}$	Clifford algebra of signature (p,q,r), p. 26
Δ	distance relation, p. 102
$\Delta(\mathcal{J})$	subgroup of PGL($\mathcal{R}, 2$), p. 120
ℓ	dual unit vector representing an oriented line, p. 13
$\ell(\cdot, \cdot)$	bilinear form corresponding to Lie's quadric, p. 74
ϵ	dual unit, p. 8
η	embedding of points in Q3GA, p. 79
γ	element of PGL($\mathcal{R}, 2$), p. 109
$\Gamma(\mathcal{C}\ell_{(p,q,r)})$	Clifford group, p. 33
$\Gamma(\mathcal{R})$	subgroup of PGL($\mathcal{R}, 2$), p. 105
$\hat{\mu}$	extended Klein mapping, p. 21
ι	morphism of distance spaces, p. 105
$\langle \cdot, \cdot \rangle_\epsilon$	standard scalar product for dual vectors, p. 9
$[\cdot]_m$	grade-m part of a Clifford algebra element, $m \in \mathbb{N}$, p. 31
$[A^+], [B^-]$	matrix representations of geometric product, p. 36
$\mathbf{i}, \mathbf{j}, \mathbf{k}$	quaternion units, p. 6
$\mathcal{N}(\mathcal{R})$	zero divisors of the ring $\mathcal{R}$, p. 111
$\mathcal{C}(\mathcal{R})$	center of a ring $\mathcal{R}$, p. 29
$\mathcal{F}$	absolute figure of CK-space, p. 182
$\mathcal{G}$	generator on Study's quadric, p. 150

List of Symbols

$\mathcal{J}$	Jordan-system, p. 119
$\mathcal{K}$	field, p. 102
$\mathcal{M}$	module, p. 77
$\mathcal{N}^1$	null vectors of a Clifford algebra, p. 128
$\mathcal{O}$	dual orthogonal matrix, p. 13
$\mathcal{R}^\times$	set of units of a ring $\mathcal{R}^\times$, p. 33
$\mathcal{V}$	projective variety, p. 199
$\mathbb{A}$	split-complex numbers, p. 117
$\mathbb{C}$	complex numbers, p. 117
$\mathbb{D}$	ring of dual numbers, p. 8
$\mathbb{D}^n$	n-dimensional module over $\mathbb{D}$, p. 9
$\mathbb{E}^n$	n-dimensional Euclidean space, p. 7
$\mathbb{H}$	skew field of quaternions, p. 6
$\mathbb{H}_d$	dual unit quaternions, p. 10
$\mathbb{NI}(\mathfrak{A})$	inner product null space of a k-blade $\mathfrak{A} \in \bigwedge^k V$, p. 32
$\mathbb{NI}_G(\mathfrak{A})$	geometric inner product null space of a k-blade $\mathfrak{A} \in \bigwedge^k V$, p. 44
$\mathbb{NO}(\mathfrak{A})$	outer product null space of a k-blade $\mathfrak{A} \in \bigwedge^k V$, p. 31
$\mathbb{NO}_G(\mathfrak{A})$	geometric outer product null space of a k-blade $\mathfrak{A} \in \bigwedge^k V$, p. 44
$\mathbb{P}^1(\mathcal{J})$	projective line over the Jordan-system $\mathcal{J}$, p. 120
$\mathbb{P}^1(\mathcal{R})$	projective line over the ring $\mathcal{R}$, p. 103
$\mathbb{P}^n(\mathbb{R})$	n-dimensional real projective space, p. 5
$\mathbb{U}_d$	group of dual unit quaternions, p. 10
$\mathfrak{A}$	blade in $\mathcal{C}\ell_{(p,q,r)}$, p. 31
$\mathfrak{a}$	grade-1 elements resp. vectors in $\mathcal{C}\ell_{(p,q,r)}$, p. 30
$\mathfrak{C}(\mathcal{L}, \mathcal{R})$	set of chains in $\mathbb{P}^1(\mathcal{R})$, p. 106
$\mathfrak{g}$	element of the Clifford group, p. 33
$\mathfrak{I}$	pseudoscalar, p. 46
$\mathfrak{Q}$	projective quadric, p. 115
$\mathfrak{Q}^*$	non-double points of $\mathfrak{Q}$, p. 116
$\text{GL}(\mathcal{R}, 2)$	general linear group over a ring $\mathcal{R}$, p. 102
I	identity matrix, p. 13
$\text{PGL}(\mathcal{R}, 2)$	projective linear group over $\mathbb{P}^1(\mathcal{R})$, p. 104
Q	quadratic form, p. 26
SE(3)	group of spatial displacements, p. 21

List of Symbols

SO(3)	group of special orthogonal 3×3 matrices, p. 6
T(2)	additive group of planar translations, p. 92
$\Omega(\cdot,\cdot)$	bilinear form corresponding to a quadric, p. 16
ω_ϵ	dual angle, p. 14
$\parallel$	parallel relation, p. 102
$\text{Pin}_{(p,q,r)}$	Pin group, p. 34
$\Sigma(\mathcal{K},\mathcal{R},\mathcal{J})$	chain geometry over $(\mathcal{K},\mathcal{R},\mathcal{J})$, p. 120
$\Sigma(\mathcal{L},\mathcal{R})$	chain geometry over the $\mathcal{L}$-algebra $\mathcal{R}$, p. 106
$\text{Spin}_{(p,q,r)}$	Spin group, p. 35
$\varkappa$	map of Blaschke and Grünwald, p. 186
ζ	Study's kinematic mapping, p. 21
A^{n_1}	absolute n_1-subspace, p. 182
$c_{\mathfrak{abc}}$	chain defined by $\mathfrak{a},\mathfrak{b},\mathfrak{c}$, p. 129
$cr(a,b,c,d)$	cross ratio of a,b,c and d, p. 1
e_+, e_-	generators of CGA, p. 43
e_i	generators of a Clifford algebra, p. 26
J	Poincaré duality, p. 41
L_1^{n+1}	Lie's quadric ($n$dimensional), p. 74
$N(\mathfrak{a})$	norm of $\mathfrak{a}$, p. 33
N_2^6	singular quadric in $\mathbb{P}^7(\mathbb{R})$, p. 185
$O(p,q,r)$	orthogonal group, p. 34
P_I	image of identity $e \in \text{SE}(3)$ under Study mapping, p. 23
$Q_{r,q}^n$	quadric of rank r, index q and dimension n, p. 182
S_2^6	Study's quadric, p. 21
$S_\mathbb{D}^2$	Study's sphere, p. 25
$tr(\cdot)$	trace of an algebra element, p. 197
V	vector space, p. 26
V^3	the exceptional generator, p. 21
$v_\epsilon = v + \epsilon\bar{v}$	small bold Latin letters denote dual vectors, with real part v and dual part $\bar{v}$, p. 9
V_k^d	algebraic variety of dimension d and degree k, p. 17
$\bigwedge^k V$	grade-k subspace of the exterior algebra corresponding to V, p. 28

Acknowledgment

Foremost, I would like to express my deepest thanks to my two supervisors, Professor Dr. Gunter Weiss and Prof. Dr. Marco Hamann. Their patience, encouragement, and immense knowledge were key motivations throughout my PhD. They persuasively conveyed an interest in my work, and I am grateful to my advisers.

Prof. Dr. Gunter Weiss has been my supervisor through three and a half years of geometric research. I am truly thankful for his steadfast integrity, and selfless dedication to both my personal and academic development. I cannot think of a better supervisor to have. Prof. Dr. Marco Hamann is a mentor, from whom I have learnt the vital skill of disciplined critical thinking. His forensic scrutiny of my work has been invaluable. He has always found the time to propose consistently excellent improvements. I owe a great debt of gratitude to Prof. Dr. Gunter Weiss and Prof. Dr. Marco Hamann.

I would like to thank PD Dr. Boris Odehnal for offering thorough and excellent feedback on an earlier version of this thesis. In addition, a " thank you " to Prof. Dr. Brehm. He gave feedback on research, and suggested general improvements to my models. Furthermore, I thank Prof. Dr. Andrea Blunck and Prof. Dr. Hans Havlicek. Both answered me questions on chain geometry during email correspondences. A special mention for Prof. Dr. Daniel Lordick, for our fruitful discussions.

I thank my parents Manuela Klawitter and Burghard Klawitter for the encouragement during the last years.

And last but not least, I thank my girlfriend Sarah Kunte. Her patience and her emotional support did always motivate me to go further.

This work was supported by the research project " Line Geometry for Lightweight Structures ", funded by the DFG (German Research Foundation) as part of the SPP 1542.

Bibliography

[1] ABŁAMOWICZ, R., AND FAUSER, B. Mathematics of CLIFFORD - a Maple Package for Clifford and Grassmann Algebras. *Advances in Applied Clifford Algebras 15*, 2 (2005), 157–181.

[2] ABŁAMOWICZ, R., AND FAUSER, B. CLIFFORD/Bigebra, A Maple Package for Clifford (Co)Algebra Computations: ©1996-2011, RA&BF. *Available at http://www.math.tntech.edu/rafal/* (2011).

[3] ABŁAMOWICZ, R., AND SOBCZYK, G. *Lectures on Clifford (geometric) algebras and applications.* Birkhäuser, Boston, 2004.

[4] BECK, H. Über lineare Somenmannigfaltigkeiten. *Mathematische Annalen, 81* (1920), 187–218.

[5] BENZ, W. *Vorlesungen über Geometrie der Algebren.* Springer, Berlin, 1973.

[6] BLASCHKE, W. Euklidische Kinematik und nichteuklidische Geometrie. *Zeitschrift für angewandte Mathematik und Physik, 60* (1911), 61–91 und 203–204.

[7] BLASCHKE, W. *Kinematik und Quaternionen.* Deutscher Verlag der Wissenschaften, Berlin, 1960.

[8] BLASCHKE, W., AND THOMSEN, G. *Vorlesungen über Differentialgeometrie und Geometrische Grundlagen von Einsteins Relativitätstheorie, Bd. 3: Differentialgeometrie der Kreise und Kugeln.* Springer, Berlin, 1929.

[9] BLUNCK, A. A quadric model for Klingenberg chain spaces. *Geometriae Dedicata, 55* (1995), 237–246.

[10] BLUNCK, A. Chain spaces via Clifford algebras. *Monatshefte für Mathematik*, 123 (1997), 93–107.

[11] BLUNCK, A., AND HERZER, A. *Kettengeometrien: Eine Einführung.* Shaker, Aachen, 2005.

[12] BLUNCK, A., AND STROPPEL, M. Klingenberg Chain Spaces. *Abhandlungen aus dem Mathematischen Seminar der Universität Hamburg*, 65 (1995), 225–238.

[13] BOTTEMA, O., AND ROTH, B. *Theoretical kinematics.* Dover Publ, New York NY, 1990.

[14] CECIL, T. E. *Lie Sphere Geometry*, 2 ed. Springer, New York, 2008.

[15] DERELI, T., KOÇAK, Ş., AND LIMONCU, M. Degenerate Spin Groups as Semi-Direct Products. *Advances in Applied Clifford Algebras*, 20 (2010), 565–573.

[16] DORAN, C., HESTENES, D., SOMMEN, F., AND VAN ACKER, N. Lie groups as spin groups. *J. Math. Phys. 34*, 8 (1993), 3642–3669.

[17] DORAN, C. J. L., AND LASENBY, A. *Geometric Algebra for Physicists.* Cambridge University Press, 2003.

[18] DORST, L., AND FONTIJNE, D. 3D Euclidean Geometry Through Conformal Geometric Algebra (a GAViewer tutorial).

[19] DORST, L., FONTIJNE, D., AND MANN, S. *Geometric Algebra for Computer Science - An Object-oriented Approach to Geometry*, 2 ed. Morgan Kaufmann, 978-0123749420.

[20] FONTIJNE, D. *Efficient Implementation of Geometric Algebra.* PhD thesis.

[21] GALLIER, J. Clifford Algebras, Clifford Groups and a Generalization of the Quaternions The Pin and Spin Groups.

[22] GARLING, D. J. H. *Clifford algebras: An Introduction.* Cambridge University Press, Cambridge and New York, 2011.

[23] GFRERRER, A. Study's Kinematic mapping - a tool for motion design. *Recent Advances in Robot Kinematics*, Eds.: Lenarcic, J., Stanisic, M.M., Kluwer Acad. Publ., Dordrecht - Boston - London, 7-16 (2000).

[24] GFRERRER, A. Study's Kinematic Mapping, unpublished manuscript. *Institute of Geometry, Graz University of Technology, Austria.*

[25] GIERING, O. *Vorlesungen über höhere Geometrie.* Vieweg & Sohn, Braunschweig, 1982.

[26] GRÜNWALD, J. Ein Abbildungsprinzip, welches die ebene Geometrie und Kinematik mit der räumlichen Geometrie verknüpft. *Sitzungsberichte Akad. Wiss. Wien*, 12 (1911), 677–741.

[27] GUNN, C. *Proceedings of AGACSE 2010: On the Homogeneous Model of Euclidean Geometry.* Springer, 2011.

[28] GUNN, C. *Geometry, Kinematics, and Rigid Body Mechanics in Cayley-Klein Geometries.* PhD thesis, Technical University Berlin, Berlin, December 9, 2011.

[29] HAACK, W. *Differential-Geometrie Teil 1 und 2.* Wolfenbüttel Verlagsanstalt G.m.b.H., Wolfenbüttel and Hannover, 1948.

[30] HAMANN, M. Line-symmetric motions with respect to reguli. *Mechanism and Machine Theory 46*, 7 (2011), 960–974.

[31] HAVLICEK, H. *Lineare Algebra für Technische Mathematiker.* Heldermann, Lemgo, 2006.

[32] HAVLICEK, H. From pentacyclic coordinates to chain geometries, and back. *Mitt. Math. Ges. Hamburg*, 26 (2007), 75–94.

[33] HERZER, A. *Handbook of Incidence Geometry: Chain Geometries.* Elsevier Science, 1995.

[34] HESTENES, D. The Design of Linear Algebra and Geometry. *Acta Applicandae Mathematicae*, 23 (1991).

[35] HESTENES, D. Old Wine in new Bottles - A new algebraic framework for computational geometry. *E. Bayro-Corrochano & G. Sobczyk (eds), Advances in Geometric Algebra with Applications in Science and Engineering* (2001), 1–14.

[36] HESTENES, D., AND SOBCZYK, G. *Clifford Algebra to Geometric Calculus: A unified Language for Mathematics and Physics*. D. Reidel, Dordrecht and Boston, 1984.

[37] HOTJE, H. Zur Einbettung von Kettengeometrien in projektive Räume. *Math. Zeitschr.*, 151 (1976), 5–17.

[38] HUSTY, M., KARGER, A., SACHS, H., AND STEINHILPER, W. *Kinematik und Robotik*. Springer, Berlin [u.a.], 1997.

[39] KARZEL, H. Kinematische Algebren und ihre geometrischen Ableitungen. *Abhandlungen aus dem Mathematischen Seminar der Universität Hamburg*, 1 (1974), 158–171.

[40] KLAWITTER, D. A Clifford algebraic Approach to Line Geometry. *Advances in Applied Clifford Algebras* (2014).

[41] KLAWITTER, D., AND HAGEMANN, M. Kinematic mappings for Cayley-Klein geometries via Clifford algebras. *Beiträge zur Algebra und Geometrie / Contributions to Algebra and Geometry 54*, 2 (2013), 737–761.

[42] KLEIN, F. Ueber Liniengeometrie und metrische Geometrie. *Mathematische Annalen*, 5 (1872), 106–126.

[43] KLEIN, F., AND BLASCHKE, W. *Vorlesungen über höhere Geometrie*. Springer, Berlin, 1926.

[44] KOWOL, G. *Projektive Geometrie und Cayley-Klein Geometrien der Ebene*. Birkhäuser, Basel, 2009.

[45] LI, H., HESTENES, D., AND ROCKWOOD, A. A Universal Model for Conformal Geometries of Euclidean, Spherical and Double-Hyperbolic Spaces. *Geometric Computing with Clifford Algebras* (2001), 77–99.

[46] LI, H., HESTENES, D., AND ROCKWOOD, A. Generalized Homogeneous Coordinates for Computational Geometry. *Geometric Computing with Clifford Algebras* (2001), 27–52.

[47] LI, H., AND ZHANG, L. Line geometry in terms of the null geometric algebra over $\mathbb{R}(3,3)$, and application to the inverse singularity analysis of generalized stewart platforms. In *Guide to Geometric Algebra in Practice*, L. Dorst and J. Lasenby, Eds. Springer London, 2011, pp. 253–272.

[48] MACDONALD, A. A Survey of Geometric Algebra and Geometric Calculus, 2008.

[49] MCCARTHY, J., AND AHLERS, S. Dimensional synthesis of robots using a double quaternion.

[50] MCCARTHY, J. M. *An Introduction to Theoretical Kinematics*. MIT Press, Cambridge and Mass, 1990.

[51] MÜLLER, E. *Vorlesungen über Darstellende Geometrie: Konstruktive Behandlung der Regelflächen bearb. von Josef Leopold Krames*, vol. 3. Deuticke, Leipzig and Wien, 1931.

[52] ONISHCHIK, A., AND SULANKE, R. *Projektive und Cayley-Kleinsche Geometrie*. Springer, 2008.

[53] PERWASS, C. B. U. *Geometric Algebra with Applications in Engineering*. Springer, Berlin and Heidelberg, 2009.

[54] PFURNER, M. *Analysis of spatial serial manipulators using kinematic mapping*. PhD thesis, University of Innsbruck, Austria, Innsbruck.

[55] PORTEOUS, I. R. *Clifford Algebras and the Classical Groups*. Cambridge University Press, Cambridge, 2009.

[56] POTTMANN, H., AND WALLNER, J. *Computational Line Geometry*. Springer, Berlin and New York, 2001.

[57] RAVANI, B., AND SPROTT, K. Kinematic Generation of Ruled Surfaces. *Advances in Computational Mathematics*, 17 (2002), 115–133.

[58] SCHRÖCKER, H.-P., AND JÜTTLER, B. The Simplest Truly Spatial Motion. *Conference on Geometry: Theory and Applications, Pilsen* (June 29, 2009).

[59] SELIG, J. M. *Geometric Fundamentals of Robotics*, 2nd ed. Springer, New York, 2005.

[60] STRUVE, H., AND STRUVE, R. Projective spaces with Cayley-Klein metrics. *Journal of Geometry 81*, 1-2 (2004), 155–167.

[61] STUDY, E. *Geometrie der Dynamen*. B.G. Teubner, 1903.

[62] VELDKAMP, G. R. On the use of dual numbers, vectors and matrices in instantanoues, spatial kinematics. *Mech. Mach. Theory*, 11 (1976), 141–156.

[63] WALTER, D. R., AND HUSTY, M. On implicitization of Kinematic Constraint Equations.

[64] WANG, W., AND JOE, B. Interpolation on quadric surfaces with rational quadratic spline curves. *Computer Aided Geometric Design*, 14 (1997), 207–230.

[65] WEISS, G. Vorlesungen aus Liniengeometrie, unpublished manuscript. *Institute of Geometry, Vienna University of Technology*.

[66] WEISS, G. Zur euklidischen Liniengeometrie I-IV. *Sitzungsberichte der Österreichischen Akademie der Wissenschaften 1978-1982*.

[67] WIKIPEDIA. Benz plane, http://en.wikipedia.org/wiki/benz_planes. *Wikipedia, The Free Encyclopedia* (April, 2014).

[68] ZAMORA-ESQUIVEL, J. G(6,3) Geometric Algebra. *9th International Conference on Clifford Algebras and their Applications in Mathematical Physics, Weimar* (15-20 July 2011).

MIX
Papier aus verantwortungsvollen Quellen
Paper from responsible sources
FSC® C105338

If you have any concerns about our products,
you can contact us on
ProductSafety@springernature.com

In case Publisher is established outside the EU,
the EU authorized representative is:
**Springer Nature Customer Service Center GmbH
Europaplatz 3, 69115 Heidelberg, Germany**

Printed by Libri Plureos GmbH
in Hamburg, Germany